Historical plant geography

By the same author:

(Edited with A. C. Jermy) *Chalk grassland: studies on its conservation and management* (Kent Trust for Nature Conservation)

(Edited) *Nature and Man in South East Asia* (School of Oriental and African Studies)

Historical plant geography
An introduction

Philip Stott

Lecturer in Geography,
School of Oriental and African Studies,
University of London

London
GEORGE ALLEN & UNWIN
Boston   Sydney

First published in 1981

GEORGE ALLEN & UNWIN LTD
40 Museum Street, London WC1A 1LU

© P. A. Stott 1981

British Library Cataloguing in Publication Data

Stott, Philip Anthony
   Historical plant geography.
   1. Phytogeography
   I. Title
   581.9   QK101   80-41627

   ISBN 0-04-580010-3
   ISBN 0-04-580011-1 Pbk

Typeset in 10 on 12 point Bembo by Bedford Typesetters Ltd
and printed and bound in Great Britain by
William Clowes (Beccles) Limited, Beccles and London.

Preface

Unlike so many plant geography and biogeography books published during the last decade, this book is neither about the study of vegetation nor the concept of the ecosystem. In writing it I have had in mind a very different tradition, one which above all concerns itself with the study of the geographical distribution of individual plant species and natural plant groups over the surface of the globe. This is the subject which has long been known as historical plant geography, a term probably introduced by Stromeyer in 1800 and first given wide currency by Schouw in 1823. It eventually became the undisputed name for the subject with a classic work by the Russian plant geographer, Wulff, which was published in English in 1943 and entitled *An introduction to historical plant geography*. Although the term can be objected to on the grounds that the argument employed is not always based on the interpretation of genuine 'historical evidence', such as the fossil record, but frequently depends on deduced past events, I have continued to use it in this book for the sake of tradition and continuity.

The book is presented primarily as an introduction to historical plant geography for undergraduate students in university and polytechnic departments of geography, biology, botany, ecology and environmental sciences. It should also prove suitable for students in colleges of education and for sixth formers with an interest in biogeography. In no way, however, is the book a substitute for the great works on plant geography discussed in Chapter 1 and listed in the Bibliography. Any serious student will have to move on from this introduction if he wishes to pursue the subject in depth. The basic aim of the book is to provide an easily understood guide to some of the more important theories and problems.

From my own teaching experience, I have come to the conclusion that such a guide is much needed. Sadly, many students are very easily daunted by the sheer size and thoroughness of most of the major works on the subject, certain of which are now also somewhat dated. Moreover, whereas students in departments of geography frequently lack a training in biology, those in biology departments often have little knowledge of subjects like cartography and geotectonics. Sections of this book have been written specifically to help these groups of people. I have also found it necessary throughout the book to emphasise the crucial link that must exist between the subjects of historical plant geography and taxonomy, and I would strongly recommend that all readers follow up at least the additional reading indicated for Chapters 2 and 9.

Students who lack a basic knowledge of the plant kingdom should consult the introductory survey by Bold (1977) and, for flowering plants in particular, the beautifully illustrated publication edited by Heywood (1978). Reference should also be made to Willis (1973). Finally, if the book is used as a basis for teaching, I should like to suggest that it is supplemented by the provision of detailed examples and case studies of plant distributions, which can then be used to illustrate and develop the theories discussed.

PHILIP STOTT
March 1980

References

Bold, H. C. 1977. *The plant kingdom,* 4th edn. Englewood Cliffs, NJ: Prentice-Hall.

Heywood, V. H. (ed.) 1978. *Flowering plants of the world.* Oxford: Oxford University Press.

Schouw, J. F. 1823. *Grundzüge einer allgemeinen Pflanzengeographie.* Berlin: G. Reimer.

Stromeyer, F. 1800. *Commentatio inauguralis sistens historiae vegetabilium geographicae specimen.* Göttingen: H. Dieterich.

Willis, J. C. 1973. *A dictionary of the flowering plants and ferns,* 8th edn (revised by H. K. Airy Shaw). Cambridge: Cambridge University Press.

Wulff, E. V. 1943. *An introduction to historical plant geography* (trans. from Russian by E. Brissenden). Waltham, Mass.: Chronica Botanica.

Acknowledgements

My chief debt is to my former teacher, Dr Francis Rose, of the Department of Geography, King's College London, who first introduced me to E. V. Wulff's great work, *An introduction to historical plant geography,* which so stimulated my early interest in the subject. Since then, I owe a great deal to the encouragement and help given to me by many colleagues, but especially to my Head of Department, Professor C. A. Fisher, and to Professor B. W. Hodder and Mr J. Sargent, who have recently been in the position of Acting Heads. A number of experts have been kind enough to read and comment on parts of the typescript; and for this I am particularly grateful to Mr R. W. Bailey, Dr J. R. Flenley, Professor R. W. Holm, Dr P. D. Moore, Dr N. K. B. Robson, Dr J. D. Sauer, Dr D. W. Shimwell and Dr E. M. Yates. To Dr Robson, as well as to Dr W. T. Stearn, I am deeply indebted for help with Chapter 2. Throughout the book, however, responsibility for any errors of fact or interpretation remains mine alone.

Much of the work of typing was undertaken by Mrs E. Bradshaw and Mrs M. Swiny, to whom I am very grateful. The rest was accomplished on borrowed typewriters (for which much thanks), my own having been involuntarily donated to British Rail. Mr P. Fox, photographer of the School of Oriental and African Studies, kindly helped with the preparation of photographic material. I should also like to express my warmest thanks to Mr R. Jones and Mr G. D. Palmer of George Allen & Unwin for their constant encouragement and unstinting assistance. Finally, I am indebted to my wife, Anne, for her patience and understanding, which eventually outweighed the efforts of my children to halt the book's far from inexorable progress.

For their kind permission in allowing the use of illustrative and other material, I should like to thank the following individuals and organisations:

Botanical Society of the British Isles (1.1);* Controller of Her Majesty's Stationery Office and the Director and Librarian, Royal Botanic Gardens, Kew (1.2, 2.1); the *Illustrated London News* (2.2); Dr N. K. B. Robson and the Trustees of the British Museum (Natural History) (2.3 & 2.4); Mrs Supee S. Larsen (Botanical Institute, University of Aarhus) and the Editors of *Grana* (2.5); the Linnean Society of London (2.6, 4.2, 6.5, 7.4, 7.9, 8.3, 8.4 and Tables 7.1, 8.2); E. S. Edees (2.7); J. Heath and the Biological Records Centre, Institute of Terrestrial Ecology (3.4); Executive Director of the American Congress on Surveying and Mapping and John Wiley (3.5); Head of Department, Spermatophytes–Pteridophytes, Jardin botanique national de Belgique (3.7); Managing Editor, *Quarterly Review of Biology* (3.8, 7.1, 7.2 &

*Figures in parentheses refer to illustrations unless otherwise stated.

7.6); Hodder & Stoughton (3.9); *Flora Malesiana* and Oxford University Press (4.3); Masson (4.4); Blackwell Scientific (5.1, 6.4 and Table 5.2); Dr J. Levitt and Academic Press (5.2); Duckworth & Co. (5.3); Collins (5.4); Cambridge University Press (5.5, 5.12); John Wiley (5.6); W. H. Freeman (5.11); Methuen (6.1); Longman Group (6.2); Dr E. C. Nelson, National Botanic Gardens, Dublin (7.7 & 7.8); Dr I. B. K. Richardson and Academic Press (8.1 and Table 8.1); Dr J. R. Flenley, Department of Geography, University of Hull (8.2); *Svensk Botanisk Tidskrift* (8.5); Prentice-Hall (Table 9.1); The Editors, *New Zealand Journal of Botany* and the New Zealand Department of Scientific and Industrial Research (9.1).

Contents

List of tables

'Les plantes ne sont pas jetées au hasard sur la terre.'

(Saint-Pierre, quoted by
Stromeyer on the title page
of *Commentatio inauguralis
sistens historiae vegetabilium
geographicae specimen*, 1800.)

1   *Plants and area*

Plant geography is the science of the distribution of plants. In this science the subject matter of one discipline, botany, is considered from the standpoint of another, geography. Not surprisingly, a shallow detractor from the subject might regard it as a case of 'plants with maps'. He would be wrong. Although the description of plant distributions rightly involves a distinct cartographic tradition, the interpretation of those distributions demands a complex synthesis of many more specific disciplines that are concerned with the geology, geography and biology of the Earth. In consequence, the detailed definition of plant geography, like that of geography itself, remains elusive.

Moreover, the content of plant geography changes with the current scientific thinking, and at no one time are all its different facets brought into focus. Unfortunately, in recent years one of the subject's more central themes – namely that concerned with the geographical distribution of plant species, genera and families – has been somewhat neglected by those who call themselves plant geographers (Stoddart 1977). With this modest book it is hoped to redress the balance a little.

Plant geography: the main traditions

The basic questions asked in plant geography are where does a plant occur and why does it occur where it does? Both are concerned with the plant's area of distribution, the one demanding a definition of this area, the other an interpretation of it. But what exactly constitutes a plant's 'area of distribution'? There are three possible answers or approaches, each with a slightly different scientific emphasis.

The first is a purely *biological* answer. For example, a root **parasite,**★ such as the toothwort (*Lathraea squamaria*†), may be said to be distributed on the roots of certain woody plants. In Great Britain *L. squamaria* is parasitic on the roots of hazel (*Corylus avellana*) and elm (*Ulmus* spp.) and such a statement constitutes a biological interpretation of its distribution. But the toothwort, and its **hosts,** form an integral part of certain vegetation associations, and their distribution may thus be further defined *ecologically*. In the *Flora of the British Isles,* Clapham, Tutin and Warburg (1962) describe the toothwort as being distributed 'in moist woods and hedgerows on good soils and locally common in some limestone areas; to 1000 ft in N. England'. Here the distribution is defined by habitat and the emphasis is ecological. But the area of distribution may also be seen in a

★Words and phrases printed in **bold** type are defined briefly in the Glossary, pp. 130–2.
†All nomenclatural authorities are given in the Index of plant names, pp. 145–8.

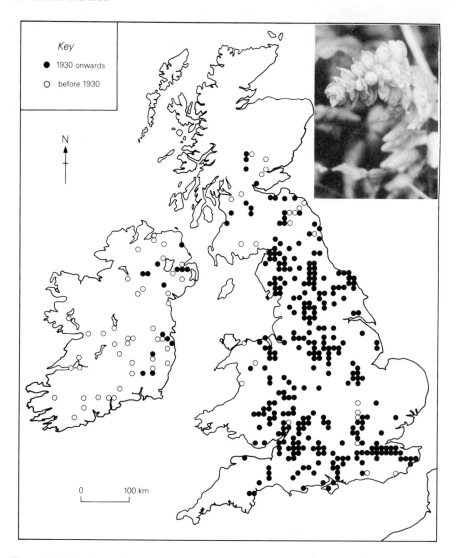

Figure 1.1   Toothwort (*Lathraea squamaria*) – its geographical distribution in the British Isles (*after* Perring & Walters 1962, 236; reproduced by kind permission of the Botanical Society of the British Isles). The inset shows toothwort on hazel (*Corylus avellana*) at Nine Horse Wood, New Ash Green, Kent in May 1975.

purely *geographical* sense. 'Throughout Europe and W. Asia to the Himalaya' is the same *Flora*'s comment on the geographical occurrence of *L. squamaria*; its geographical distribution in the British Isles is shown in Figure 1.1.

These three different approaches to the distribution of plants have all, at one time or another, played their part in the thinking of plant geographers and, of course, they are not mutually exclusive. The geographical tradition is perhaps the oldest, with a notable pre-Darwinian ancestry, founded not only on the gradual exploration, discovery and description of the world's plant life, but also on some early masterpieces of plant geography, in particular the *Essai sur la géographie des plantes* of Humboldt and Bonpland (1805, German edition 1807) and the *Géographie botanique raisonnée* of Alphonse de Candolle (1855), works which remain of more than historical interest. But it was with the evolutionary understanding of Darwin, Wallace, Huxley and, pre-eminent from the point of view of plant geography, Hooker (Turrill 1953, Fig. 1.2), that the subject began to take on its modern theoretical dimensions. The geographical distribution of living organisms, already seen as a crucial line of argument in *The Origin of Species* (1859), was now reinvestigated and nowhere more so than in Germany with the important contributions of Grisebach (1872, 1880), Engler (1879–82) and Drude (1890, French edition 1897).

Figure 1.2 Joseph Dalton Hooker – one of the founders of modern plant geography. He was among the first to recognise the change that the new evolutionary thinking of Darwin's *On the origin of species by means of natural selection* (1859) brought to the subject. (Crown copyright; reproduced with the kind permission of the Controller of Her Majesty's Stationery Office and of the Director, Royal Botanic Gardens, Kew.)

Yet, paradoxically, the new thinking also proved the bane of the geographical tradition, particularly in North America and Britain, leading to its partial demise for some forty years from the turn of the century with certain useful contributions excepted, especially those of Willis (1922), Guppy, Ridley, de Vries and Diels. The great discussion on evolution now brought into focus the relationships of organisms with their environment and the ecological tradition became established. Significantly, the word 'ecology' was first coined in 1866 by Ernst Haeckel. In his original definition the term was applied solely to the study of the relationships of one organism to its environment, a subject for which the present name is **autecology** (see below). He described ecology as 'the total science of the relationships of the organism to the surrounding environment, within which we can include in a further sense all conditions for life'. However, the earliest comprehensive plant geographies in the ecological mould tended to emphasise the study of whole vegetation formations. These were published at the end of the century with the still basic texts of Warming (1896, English edition 1909), Schimper (1898, English edition 1903) and Raunkiaer (1907, 1934), a scholar whose work on **life-forms** also exemplifies the biological tradition of the subject. Since this time the ecological tradition has gone from strength to strength, with three main lines of development.

The first of these is the autecological tradition, founded, as we have seen, on the study of the relationships between *individual* organisms and their environment (e.g. Daubenmire 1974). The second is the **synecological** approach which looks at the relationships of *groups* of organisms and the environment. This latter school of thought is now based on the concept of the **ecosystem** (a term first introduced by Tansley in 1935), but it is also apparent in the parallel, yet independent, work of Troll (1939) on *Landschaft* ecology (see Lauer & Klink 1978) and in the work of Sukachev on the **biogeocoenosis** carried out during 1944–5. These separate developments of ecological concepts, the one geographical (Troll), the other derived from forestry (Sukachev), have received relatively little examination in Anglo-American literature, although they are part of long-established fields of study (see Sukachev & Dylis 1968). The term **biocoenosis** was first introduced in 1877 by Karl Möbius to describe the internal relationships of living communities.

In the main, however, it is the concept of the ecosystem which has taken a stranglehold on the subjects of both ecology and plant geography, so much so that many books purporting to be biogeographies or plant geographies are little more than manuals on ecosystem studies (e.g. Pears 1977, Tivy 1971, Watts 1971). This restrictive view of plant geography is unfortunate for it denies not only the geographical tradition but also much of the ecological tradition itself. Autecology, for example, has not received the attention that it should.

The third line of development has been concerned with the description and classification of vegetation, which, as noted above, was an important theme implicit in the early works of Warming and Schimper. Recent reviews of this field are provided by Grieg-Smith (1978), Harrison (1971) and Shimwell (1971), among others.

Yet the geographical tradition has not been totally neglected and the post-1940 period has seen the publication of a number of important works. Perhaps the greatest of these is the *Historical plant geography* of E. V. Wulff, the first volume of which was translated into English from the original Russian in 1943. The second volume, which is in Russian only, appeared in 1944 and was probably to have been followed by a concluding third volume. Unfortunately, it seems that this was never completed, for the author was killed in the siege of

Leningrad in 1941. His extant work, however, remains one of the finest contributions ever made to the geographical tradition. Other studies, which vary greatly in their importance and value, have included the rather technical *Foundations of plant geography* by Cain (1944), *The geography of the flowering plants* by Good (first published in 1947 and now in its fourth edition, 1974) and the works of Cox *et al.* (1976), Croizat (1952, 1958, 1960), Daubenmire (1978), Gleason and Cronquist (1964), Lauer and Klink (1978), Lemée (1967), Ozenda (1964), Polunin (1960), Rothmäler (1950), Szafer (1964, English edition 1975) and Walter (1954).

However, despite this maintenance of the geographical tradition, its place in higher education has been increasingly eroded in recent years, whereas the ecological tradition has flourished and grown out of all proportion. For this reason the present book concentrates on the geographical tradition and looks at 'areas of distribution' from that point of view. We are therefore concerned with the *area geographica,* or geographical range, of plants and with that aspect of plant geography designated by Wulff as 'historical plant geography'. Speaking of zoogeography – the science of the distribution of animals – Hesse, writing in 1924, drew the now commonly accepted distinction between 'ecological' and 'historical' animal geography (Illies 1974). The same division obtains in plant geography but has sometimes been ignored or not recognised. Both traditions are vital if a comprehensive view of the science is to be maintained.

The concept of area

Historical plant geography is essentially the science of area applied to plants. Its first task is to establish the distribution of plant **taxa** (e.g. species, genera and families) in geographically defined areas, and its second is to interpret the origins and present status of these areas. The term geographical range is applied to the 'entire region of distribution or occurrence of any taxonomic unit . . . whatever its rank' (Cain 1944, 147). The term may be limited geographically and may be used to refer to the occurrence of a particular taxonomic entity (**taxon; pl. taxa**) within a local district, county, country or continent, or to its entire region of distribution at the world scale. As long as these reference points of scale are clear, the concept of area may be justifiably restricted.

When mapped, no two plant areas are exactly the same in size, shape, topography or geographical territory embraced. The word 'topography' is used here in the sense of Wulff and refers to the local character of a plant's distribution within its area as a whole. For example, although the water crowfoot (*Ranunculus aquatilis*) is found throughout the lowland parts of Great Britain and Ireland, within that general area it is confined to ponds, ditches and streams. It has a localised 'topography' within its basic distribution.

A crucial stage in all plant geographical work is the identification of recurrent patterns within the great diversity that exists. The assumption that the same form may be produced by the same process, although always a dangerous hypothesis, is fundamental to much research in plant geography. Plant geographers have, in consequence, devised a number of ways of classifying plant areas, and these are discussed in Chapter 4.

Very few areas are in any sense truly continuous except at the most local level. At the world scale, the prize for the most continuous distribution must go to the grass family, the Gramineae (Poaceae), which reaches the very borders of Antarctica and occupies the furthest land masses of the Northern Hemisphere. Moreover, it is a dominant element in many vegetation formations throughout the world, and in nearly all formations it is well-represented or at least present. Other **cosmopolitan** families ranking close to the Gramineae in their ubiquity are the Compositae (Asteraceae), the Caryophyllaceae and the Cyperaceae, with the genus *Carex* as its key temperate representative and the genus *Cyperus* found mainly in the tropics.

Some authors have argued that plant geographers should concentrate their attention on the main portions of a plant area, the parts which they term the compact area. Others, such as Cain and Hultén, rightly reject this suggestion and point out that it is more often the outlying stations of a distribution – the radiations or relicts – that provide the clues by which we may interpret the history of the area. Perhaps, however, it is always wisest to take into account the total distribution pattern and the full topography of a distribution if a true understanding is to be achieved.

A further, though necessary, division of areas is into natural areas and artificial areas. Plant geography is in the main concerned with natural areas and distributions which are the products of natural dispersal mechanisms and agencies of dispersal, excluding man. Artificial areas arise where man has a significant rôle in their formation, whether deliberate or unconscious. Natural areas become artificial when the actions of man markedly reduce or extend the area, or create a new type of distribution by changing, for example, a formerly continuous distribution into a broken or disjunct distribution. In some instances the hand of man is easy to detect; in others his contribution remains obscure. The problem of distinguishing the natural from the artificial has proved increasingly troublesome in recent years, especially with the rapid growth of biological interchange between countries. In the Preface to his latest edition of *The geography of the flowering plants* (1974), Good concludes, perhaps a little pessimistically, that 'the point has now been reached which makes it seem doubtful whether, in future, any considerable proportion of new geographical records can be accepted simply at their face value'.

A model area

Man, however, represents only one of the major disruptions to plant areas. Geological instability (with the movements of plates and continents, the building of mountains and the submergence of land), climatic change, the evolution of new and more vigorous **competitors,** disease, and animal activity all help to destroy, separate, alter or enlarge plant areas. An appreciation of these processes is essential in any interpretation of present-day distributions. Yet, for the moment, it would be helpful to imagine an isolated island, immune from such influences, unpeopled, with a stable geology, climate and ecology, and free from all outside intrusions and invasions (Fig 1.3). Bounded by a quiet sea, a great western plain runs down to graded beaches, save in the east where it is backed by high mountains and forested hills that meet the sea in rocky cliffs. On this serene island there arises a new plant species (*x*) at a locality (*O*) on the western plain.

Species *x* is a true plains species, unable to grow on hills or mountains, or by

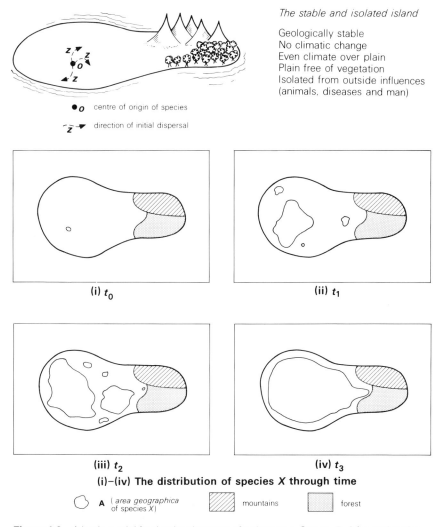

The stable and isolated island

Geologically stable
No climatic change
Even climate over plain
Plain free of vegetation
Isolated from outside influences
(animals, diseases and man)

●*O*   centre of origin of species

z⟶   direction of initial dispersal

(i) t_0

(ii) t_1

(iii) t_2

(iv) t_3

(i)–(iv) The distribution of species *X* through time

A (*area geographica* of species *X*)   mountains   forest

Figure 1.3   A basic model for the development of a plant area. See pp. 7–8 for explanation.

the sea or in the oceans. Moreover, it is a plant of open habitats, avoiding forests and wooded areas, which present a natural barrier to its spread. The development of its geographical range (A) from the time of its origin (t_0) to the time when it has colonised all the land suitable for its establishment (t_3) is suggested hypothetically in maps (i) to (iv) in Figure 1.3. Its final limits are determined by the ecological barriers of the mountains and the forests, the beaches and the sea. Unless the geological structure of the island changes, or the present climate alters, or the forest expands or contracts, or until plant x is wiped out by disease or damaged by animals, or until a new, more vigorous competitor arises or man comes to the island to clear the land and cultivate the soil, the area of plant x will remain unchanged. The boundaries of its distribution will be those mapped for time (t_3) in Figure 1.3. Plant x has achieved a continuous area with boundaries that are clearly determined by the absolute ecological barriers.

In this basic model, the distribution of plant x and the evolution of its geographical range (A) may be said to be the function of five main factors:

O = the geographical location of the centre of origin;
D = the nature and efficiency of the dispersal mechanisms and the rate at which plant x can establish itself in new localities;
P = the basic ecological requirements of plant x (its autecology);
Z = certain chance factors, e.g. the direction of initial dispersal and of subsequent dispersals from both the centre of origin (O) and later secondary centres;
t = time, defined as the age of species x, i.e. the time species x has been in existence (t_0, t_1, t_2, t_3 and so on); the older it is, the greater will be its area.

These five factors represent the internal controls (I) of a plant distribution and, where all external influences (E) are removed, it is their complex interaction which will determine the spread of a plant. But in nature this is rarely the case, for all land and sea is subject to change and the world is full of competitors and enemies. There is, therefore, no simple model for plant distribution in which an uninterrupted process leads to an inevitable conclusion determined by absolute ecological barriers. Although plant geographers must of necessity be concerned with the internal factors of the origin and age of plants, they are just as involved with the external processes of geological, climatic and habitat change, which disrupt the evolving and developing distribution patterns.

The four stages of plant geography

An understanding of process, however, is the final goal of plant geography, but this fourth stage must be preceded by three earlier stages if its theories and explanations are to rest on firm foundations. These stages are not strictly historical, although they are reflected to some extent in the historical development of the subject from the 17th century onwards. Basically, the stages should be seen as the essential theoretical structure of the plant geographer's discipline, representing the main steps of intellectual activity that are necessary for the successful completion of any plant geographical work. Moreover, it should be recognised that it is rare for an individual research worker to be fully involved in all four stages.

The first stage, which is described in Chapter 2, is perhaps the most important of all. It involves the collection, identification and recording of plants in the field. This stage provides the basic distributional information for all plant geographical work. If the methods used are not scientifically rigorous, the whole subject will rest on unsure foundations. Chapter 2 could therefore be regarded as the most fundamental in this book, for it seeks to analyse how plant geographers safeguard both their taxonomic and their site information.

The second stage is the mapping of plant distributions using the information collected in the first stage of the work. The quality and scientific integrity of plant distributional maps is considered in Chapter 3. Again this is a basic stage which can do much to mar or enhance eventual theories and explanations.

The third stage is the classification of the plant distributions that have been mapped into recognisable patterns or groupings. Although there are nearly as many different classifications and classificatory approaches as plant geographers, a number of standard methods have been developed and these are discussed in Chapter 4.

The fourth, and final, stage of plant geography is the generation and testing of theories to explain the types of distribution that have been discovered and described in the first three stages. The acceptability of these theories will above all else depend on the care that has been exercised in the three preceding stages. Although it is this last stage that most excites the mind, it is of little value if the previous work has been neglected, skimped or mishandled. The theories of plant geography, which are discussed in the second part of this book, are only as good as the collections, records, identifications, maps and classifications on which they are based.

Further reading

Useful introductory texts are marked with an asterisk.

General

*Cox, C. B., I. N. Healey and P. D. Moore 1976. *Biogeography: an ecological and evolutionary approach*, 2nd edn. Oxford: Blackwell Scientific.

Fosberg, F. R. 1976. Geography, ecology and biogeography. *Ann. Assoc. Am. Geogrs* **66**, 117–28.

J. Biogeog., 1974 onwards.

Stoddart, D. R. 1977. Biogeography. *Progress Phys. Geog.* **1**, 537–43.

Stoddart, D. R. 1978. Biogeography. *Progress Phys. Geog.* **2**, 514–28.

Zoogeography

*Illies, J. 1974. *Introduction to zoogeography*. London: Macmillan.

Historical plant geography

Cain, S. A. 1944. *Foundations of plant geography*. New York: Harper & Row. (Reprinted 1974. New York: Hafner.)

Good, R. 1974. *The geography of the flowering plants*, 4th edn. London: Longman. (First published 1947.)

Szafer, W. 1975. *General plant geography*. Springfield, Virginia: U.S. Department of Commerce. (English trans. by H. M. Massey of *Ogólna geografia roślin*, 3rd edn. Warsaw: Państwowe Wydawnictwo Naukowe, 1964.) (See particularly 185–303.)

Turrill, W. B. 1953. *Pioneer plant geography: the phytogeographical researches of Sir Joseph Dalton Hooker*. The Hague: Nijhoff.

Turrill, W. B. 1963. *Joseph Dalton Hooker. Botanist, explorer and administrator.* London: Nelson.

Wulff, E. V. 1943. *An introduction to historical plant geography* (Trans. from Russian by E. Brissenden). Waltham, Mass.: Chronica Botanica. (Although now dated and increasingly difficult to find, this remains one of the finest works on the theory of historical plant geography.)

Ecology

*Daubenmire, R. F. 1974. *Plants and environment. A textbook of autecology*, 3rd edn. New York: John Wiley. (First published 1947.)

*Kormondy, E. J. 1976. *Concepts of ecology*, 2nd edn. Englewood Cliffs, NJ: Prentice-Hall.

Lauer, W. and H. J. Klink 1978. *Pflanzengeographie*. Darmstadt: Wissenschaftliche Buchgesellschaft. (For the work of Troll on *Landschaft* ecology: also a section on historical plant geography.)

Sukachev, V. and N. Dylis 1968. *Fundamentals of forest biogeocoenology*. (English trans. from Russian by J. M. MacLennan.) Edinburgh & London: Oliver & Boyd.

I ESTABLISHING PATTERNS OF DISTRIBUTION

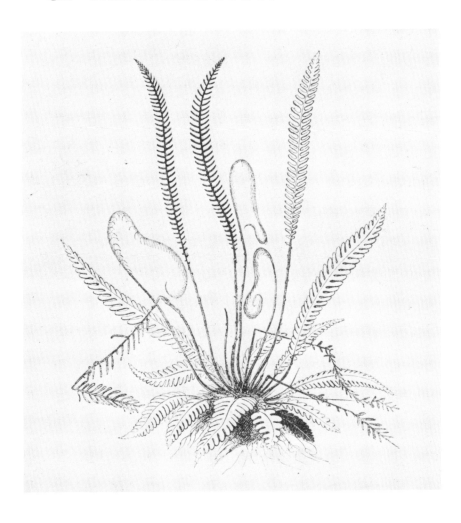

2 Plant collecting and recording: the taxonomic basis of plant geography

All plant geography must begin with the discovery of a plant in the field and rests on the correct identification and naming of this plant and the accurate recording of the place of discovery. Without the collection of specimens and the recording of localities in the field, there could be no subject 'plant geography'. Without the reliable identification of the specimens collected or noted, and the detailed geographical coordinates of the sites concerned, the maps, the classifications and the theories of plant geography have no solid foundations. It cannot be emphasised too much that any successful work in plant geography is only possible when it is based on sound taxonomy, that is on the correct identification, naming and, in some cases, phylogenetic classification of the taxa involved, and on the fullest geographical records of the sites of discovery.

Our present knowledge of the world distribution of plant families, genera and species rests on a long history of discovery – on a story which is filled with ambitious, dedicated and often bizarre characters and which relates many tales of adventure, intrigue and scientific endeavour. Man has sought out plants for many different reasons. He has been attracted by their beauty, their symbolic and 'magical' properties, their horticultural value, their medicinal properties, their potential as food plants, and yet many of the most startling discoveries have been made entirely by chance. Very rarely has the plant geographer himself seen his work from start to finish, from the collection and identification of the taxa in question, through their mapping and the classification of their distributions, to the development of theories to explain these distributions. Plant geography is essentially a secondary subject which relies on information from many sources to reach its conclusions. In this simple fact lies one of the great weaknesses of the subject for plant geographers must be continually asking two questions: how reliable is this record and was the plant correctly identified?

In many cases, without actually revisiting the localities involved, it may prove impossible to answer these questions satisfactorily, and in certain instances, for example where specimens have been destroyed or the plant is now extinct, it may not be possible at all. Because of such limitations and because so few regions of the world have been thoroughly worked by specialists, it is sometimes surprising that any plant geographer ever feels really confident that the plant distributions with which he is working are in any way reliable.

It is the purpose of this chapter to illustrate these fundamental problems more fully and to point to the main ways in which the reliability of the information needed by the plant geographer is safeguarded. Particular emphasis is laid on the need for an acceptable taxonomic basis for plant geography, and for as many records as possible to be supported by fully labelled and well curated herbarium specimens. For a fascinating and readable history of plant hunting itself, the student is referred to Whittle (1970), in which the saga of plant discovery is surveyed and the techniques of plant collecting outlined.

Yet it must not be thought that plant hunting is nothing but a precursor to the subject for it must always remain the first step in all plant geography, and the advanced systems of survey and recording current today are no more than sophisticated plant hunting, a finer net being cast over smaller areas. Even today the difficulties of plant collection and transport should not be underestimated but, in 1819, when Dr John Livingstone, then Chief Surgeon in Canton, wrote to the Horticultural Society of London on the problems of transferring plants from China, the problems were legion. What with the actual cost of shipping and the effects of varying climates, humid storage quarters, sea spray, idle and ignorant sailors, insects and rats, it is a wonder that any plants at all arrived in the botanical gardens and herbaria of Europe. Livingstone made this point forcibly in his letter:

> From my observation, I am persuaded more than one thousand plants have been sent from China for one Chinese plant, which is now cultivated in England.
> Plants purchased in China, including the freight, is [*sic*] on an average, one tael each plant, or three for one pound sterling, consequently each plant now in England, must have been introduced at the enormous expense of upwards of £330.
>
> (*The Indo-Chinese Gleaner,* 1819, **2,** IX, 130)

The case of Victoria amazonica

The long and complicated story of the discovery, description, naming and successful cultivation of one particular 'vegetable wonder', the famous giant water-lily of the Amazon (*Victoria amazonica*) (Fig. 2.1), is especially illustrative of some of the problems outlined above. It is by any standards a remarkable tale, so much so that in 1974 BBC Radio 4 turned it into a most engaging broadcast. The key figure in this programme was that Victorian genius, Joseph Paxton, who was not only the first person in England to bring the plant to flower in cultivation but also was inspired by the remarkable structure of the plant's enormous leaf to design the Crystal Palace. The extraordinary strength of the leaf soon became legendary and, as a practical test of this quality, Paxton had actually placed his seven year old daughter on one, much to the delight of the *Illustrated London News,* which published a sketch of the event (Fig. 2.2), and of *Punch,* which commemorated the occasion in verse:

Figure 2.1 *Victoria amazonica,* opening flower. (Plate II from W. Fitch 1851. Victoria regia *or, illustrations of the Royal Water-Lily, in a series of figures chiefly made from specimens flowering at Syon and at Kew.* London: Reeve & Benham. Crown Copyright: reproduced with the permission of the Controller of Her Majesty's Stationery Office and of the Director, Royal Botanic Gardens, Kew.)

Figure 2.2 Miss Annie Paxton demonstrating the strength of a leaf of the giant water-lily of the Amazon. (From the *Illustrated London News* of 17 November 1849.)

On unbent leaf in fairy guise,
Reflected in the water,
Beloved, admired by hearts and eyes,
Stands Annie, Paxton's daughter.

This giant of the flowering plants was first discovered in 1801 by Thaddaeus Haenke (1761–1817), a Bohemian botanist, doctor and mineralogist, but, it is surprising to find, his discovery remained largely unknown until 1838, even though a number of other Europeans had reported seeing it during the intervening period. One of these was a Frenchman called d'Orbigny. He actually sent off specimens of its leaves and flowers to the Muséum Nationale d'Histoire Naturelle in Paris. But only one gigantic leaf survived the journey and, instead of setting the scientific world alight, it was folded in two and deposited on a shelf in the library of the museum and forgotten!

The real breakthrough, however, came on New Year's Day 1837 when the renowned German-born explorer and botanist Robert Schomburgk (1804–65), who was in charge of an expedition sponsored by the Royal Geographical Society to what was then British Guiana, arrived at a point where the river Berbice 'expanded and formed a currentless basin'. We take up the story in his own words:

> Some object on the southern extremity of this basin attracted my attention. It was impossible to form any idea what it could be, and, animating the crew to increase the rate of their paddling, shortly afterwards we were opposite the object which had raised my curiosity. A vegetable wonder! all calamities were forgotten, I felt as a botanist, and felt myself rewarded. A gigantic leaf, from 5 to 6 feet in diameter; salver-shaped, with a broad rim of light green above, and a vivid crimson below, resting upon the water. Quite in character with the wonderful leaf was the luxuriant flower, consisting of many hundred petals, passing in alternate tints from pure white to rose and pink. The smooth water was covered with them . . . .

> (*Mag. Zool. Bot.* 1838, **2,** 441)

Table 2.1 An example of a Latin description – Lindley's Latin diagnosis of the genus *Victoria* (from the *Botanical Register* 1838, **24,** Misc. Not. 13). Note, however, the discussion of the correct authority citation for this genus in the footnote on p. 17.

Victoria

Calyx campanulatus limbo 4-partito deciduo. Petala indefinita, fauce calycis inserta, exteriora patentissima, interiora incurva multò minora. Stamina plurima petaloidea, fauce calycis inserta; exteriora fertilia libera, interiora sterilia cornuta stigmatibus a tergo adnata. Ovarium inferum multiloculare; loculis polyspermis: ovulis parietalibus; stylis in campanulum sulcatum, tubum calycis vestientem connatis; stigmatibus maximis reniformibus, carnosis. Fructus campanulatus, truncatus, carnosus, intra basin capsulam gerens medio, longè rostratam, polyspermam.

Using such descriptions, as well as sketches made by Schomburgk's draughtsman, John Morrison, and also the decayed remains of the specimens sent to London, which, although meticulously packed in cases filled with salt and water, failed to survive the journey, the Professor of Botany at University College London, John Lindley, was able to draw up a scientific diagnosis of the plant. He determined it as not only a new species but also a new genus, which, being undoubtedly the genus of the queen of flowers, was named, by gracious permission, after Queen Victoria.★

The genus *Victoria* still stands today, but not so Lindley's species name of *V. regia,* for in fact there existed an earlier description of the plant by Eduard Friedrich Poeppig who had found the giant water-lily in the course of his travels up the Igaripes, a tributary of the Amazon. Poeppig's name of *Euryale amazonica* comes from 1832 thus predating those of both Schomburgk and Lindley, and so by the rule of priority (see p. 21) the correct name of the plant is *Victoria amazonica* (Poeppig) Sowerby, Sowerby being the person who, in 1850, transferred the *Euryale amazonica* of Poeppig to the new genus *Victoria*.

This complex history fully illustrates the problems attendant upon the collection, description and naming of new taxa and this was not a small, easily overlooked plant, but one of the world's vegetable marvels. The substance of plant geography rests on thousands of such stories (after all, it is estimated that there are over 300 000 species of flowering plants alone). As indicated in the footnote below★, the story of *V. amazonica* is itself still not complete. Moreover, systematists are even today divided over the question of what family the plant belongs to: some include it in the widely defined family Nymphaeaceae, while others place it in the more narrowly defined family Euryalaceae, a family of large aquatic annuals or short-lived perennial herbs containing two genera, *Euryale* and *Victoria* itself.

The herbarium specimen

It has already been argued that the most acceptable record of a plant's occurrence at a particular site is a fully labelled and well curated herbarium specimen. If the geographical coordinates are accurately recorded, along with further details about the locality at which the specimen was collected, the herbarium specimen

★The actual authority for the new genus, *Victoria,* is still a matter for some debate. It is now taken to be Schomburgk and not Lindley (Willis, 1973, 8th edn revised by Airy Shaw) and it is true that Lindley's 1838 Latin diagnosis (*Bot. Register* 1838, **24,** Misc. Note, 13) was preceded by Schomburgk's description contained in his published letter to the Botanical Society of London (*Mag. Zool. Bot.* 1838, **2,** 440–2). The letter is dated 17 October 1837. However, on 16 October 1837 the Shakespeare Press published an earlier description by Lindley, entitled *Victoria regia*. The present work is not the place to resolve this matter and more research is needed, but the case illustrates well how nomenclatural problems can persist long after the original discovery of a plant. Table 2.1 (p. 16) gives Lindley's 1838 diagnosis as an example of a Latin description.

provides unambiguous and enduring evidence of that plant's occurrence at the given site. The identification of the plant is readily checked and can be constantly reviewed in the light of more recent taxonomic research. It is a near perfect safeguard against wrong identification and later confusion. There must always remain a shadow of doubt about many written and published records un-supported by herbarium specimens.

The art of preserving herbarium specimens has been known for well over 400 years. If a living plant is dried rapidly while compressed between two flat absorbent surfaces a wide range of its botanical characters is preserved, thus enabling the detailed study of the plant at a later date. The pressing while drying prevents shrivelling. Such a dried and pressed plant is termed a herbarium specimen. There are a number of ways of storing such specimens, but the commonest involves the mounting of the dried plant on good quality sheets of paper (herbarium sheets) as shown in Figure 2.3a. The specimen should be accompanied by the following detailed information appended on the sheet: the place of collection; the name of the collector; the collection number; the date of collection; field notes on the habitat of the plant; a description of its habit when growing; and a careful record of the characters which may have disappeared on drying, such as flower colour and scent. Certain loose material or small and delicate items, like seeds, tiny fruits or paper-thin flowers, may be lost or broken if not protected and these are usually preserved in small envelopes attached to the sheet. This procedure also facilitates their extraction and detailed study at a later date. Other items, such as large and bulky fruits, may have to be stored separately from the main sheet. These are frequently kept in air-tight boxes with a glass top, like the fruit of *Dipterocarpus dyeri* shown in Figure 2.3b. It is obviously important to establish a satisfactory cross-reference system between the main sheets and such separate material.

Not all plants are susceptible to drying and pressing and those that are not may have to be stored in other ways. Some aquatics, for example, are best stored in water or preserving liquid in bottles. One such is the type specimen (see p. 21) of *Utricularia subulata* shown in Figure 2.3c. Bryophytes and lichens are dried and kept in packets like envelopes, with the information about the collection recorded on the flap.

All the above kinds of specimen are stored in herbaria and, in the main, take up remarkably little space. To the taxonomist and the plant geographer, the herbarium is a 'library' of first importance, a data source without which their task would be impossible. With careful curation the specimens may be preserved indefinitely. All they need is protection from temperature extremes, insect damage, fungal attack and damp. These conditions are readily obtained by storage in dry cupboards or cabinets placed in rooms of moderately even temperature, and by the use of insect repellants. Great care, however, must be exercised in herbaria established in tropical countries. It should go without saying that many specimens are fragile, especially the older ones, and that these must be handled gently to avoid physical damage.

In addition to the information already listed, a herbarium sheet will usually

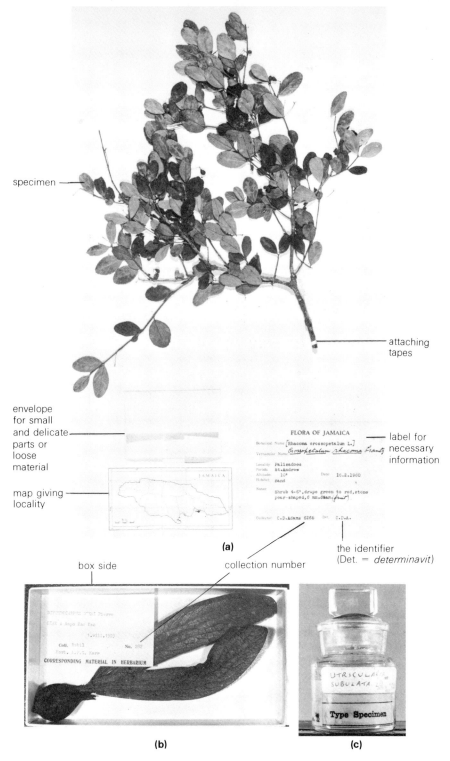

specimen —

attaching tapes

envelope for small and delicate parts or loose material

map giving locality

FLORA OF JAMAICA

Botanical Name [Rhacoma crossopetalum L.]
Vernacular Name Crossopetalum rhacoma Crantz

Locality Palisadoes
Parish St.Andrew
Altitude 10'   Date 16.2.1960
Habitat Sand

Notes Shrub 4-6',drupe green to red,stone
pear-shaped,6 mm.diam.(fruit)

Collector C.D.Adams 6266   Det C.D.A.

— label for necessary information

the identifier
(Det. = *determinavit*)

(a)

box side

collection number

CORRESPONDING MATERIAL IN HERBARIUM

UTRICULA
SUBULATA

Type Specimen

(b)

(c)

Figure 2.3   Examples of quite well prepared herbarium specimens.

carry the name of the herbarium in which it is stored, a herbarium number and a wide range of miscellaneous notes which have accumulated through time. The botanical name of the plant is nearly always present and may well have gone through a number of corrections.

Botanical nomenclature

A fully labelled and well curated herbarium specimen safeguards the basic data of the plant geographer but it does not, in itself, safeguard the scientific naming of the plant so preserved. This is the job of the International Code of Botanical Nomenclature (ICBN), which governs the scientific naming of all plants except **cultivars,** the naming of which is governed by a separate code. Only by strict adherence to these codes is it possible to maintain a truly worldwide system of nomenclature and to avoid the confusion implicit in idiosyncratic systems, with duplication, misuse and purely local application of names.

The basis of the modern method of nomenclature and classification lies in the work of the celebrated Swedish biologist, Carl Linnaeus (1707–78), effective creator of the binomial system of nomenclature for plants and animals. Under this system the name of a species comprises two parts. The first is the name of the genus (pl. genera) to which it belongs and the second is what is termed the specific epithet. Under the ICBN, these names must either be Latin names or be treated as such if coming from another language. The genus name is always a singular noun and is written with a capital initial letter, e.g. the mint genus, *Mentha.* The specific epithet, which can *only* occur in combination with a generic name, may (though not mandatory, the Code recommends this) always be written with a small initial letter, e.g. the water mint, *Mentha aquatica,* and the spearmint, *Mentha spicata.* The species name of the water mint is, therefore, *Mentha aquatica.*

The botanical name of a plant is often followed by a further piece of information which is not actually a part of the botanical name but which is a vital element in the naming process. This is known as the authority citation and is the name, usually in an abbreviated form, of the person who first published the plant name preceding it. In the case of the water mint, the citation is *Mentha aquatica* L., a well-known abbreviation showing that Linnaeus was the author of the name. The author of the name is also the authority for that name. The authority for the bee orchid, *Ophrys apifera* Huds., is, as indicated, William Hudson (1730–93). It should be noted that work is now in progress to standardise the abbreviations used.

This concept of the authority citation is crucial in the safeguarding of the scientific names of plants. Under the ICBN, for a scientific name to be valid it must be both effectively and validly published. This means that the name must be published in printed matter which is made available to the public by sale or exchange and it must be accompanied by a description of the plant to which the name is applied. This description was usually in Latin (see Table 2.1) and since

1935, in the case of higher plants, Latin has been mandatory. Names not so published are invalid and must be ignored. The author of the name is the person who first published the name according to the above rules and it is his personal name which becomes the authority citation. This citation is a ready source of information for anybody who wishes to check the original description in that it refers them to the author in whose work the name orginated.

There is, however, one further bridge that needs to be built. This is the bridge that links the publication and description of the name of a plant with the plant itself. This problem is dealt with by a system of typification. Since 1958 the name of a new family, genus or species is not accepted as validly published unless its nomenclatural type is indicated.

The type of a name is the 'element' on which the description validating the publication of that name was based. In the case of a species, this is normally a herbarium specimen. Figure 2.4 shows the type specimen of the name *Magnolia paenetalauma* Dandy. In addition to the usual material to be found on a herbarium sheet, this particular sheet also carries the reference to the original description, based on this very specimen, which was published by Dandy to validate the name in the *Journal of Botany* 1930, **LXVIII,** 206. It is worth noting that it validates the *name*. The type is simply 'a pin by which the name is attached to the plant'. Only names have types; species, genera and families do not. A type specimen under no circumstances demonstrates all the range of variation that may exist in a given taxon. In the case of a genus, the type of the name is the species on which the original description validating the name was based, and for the family, it is the genus on which the name was based. For a fuller explanation of the working of the type method, the reader is referred to the easily understood analysis in Jeffrey (1968) and to Jones and Luchsinger (1979).★

Two further matters need a brief elucidation, however. The first of these is the principle of priority, which states that when two or more names compete for the same taxon, the name that was published first, in other words the oldest name, is correct. Referring to the earlier discussion of *Victoria amazonica* (p. 17), it is this principle which necessitated Lindley's name of *V. regia,* dating from

---

★It should be noted, however, that it is necessary to recognise a number of different categories of 'types'. The most important of these are:

holotype   – the one specimen or other element used by an author or designated by him as the nomenclatural type;

isotype   – a duplicate of the holotype (part of the single gathering made by a collector at the one time) – it is always a specimen;

lectotype   – a specimen or other element selected from the *original* material to serve as a no-menclatural type where the holotype is missing or when no holotype was designated at the time of publication;

neotype   – this is cited when all the original specimens and their duplicates have been lost or destroyed;

paratype   – a specimen cited with the original description other than the holotype or isotype;

syntype   – one of two or more specimens cited by an author when no holotype was designated, or one of two or more specimens originally designated as types.

The reader is referred to Radford *et al.* (1974, Ch. 1–4) for further details.

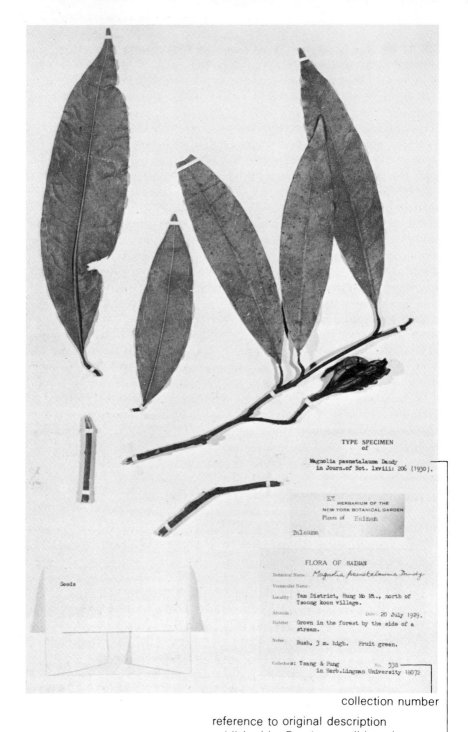

collection number

reference to original description
published by Dandy to validate the name

Figure 2.4 The type specimen of the name *Magnolia paenetalauma* Dandy.

1838, to succumb to the name *V. amazonica,* which is based on Poeppig's name *Euryale amazonica,* dating from 1832. The name *V. regia* is relegated to what is known as the synonymy of the species, which comprises a list of the different names that have been given to the species, the oldest of which is now the correct name.

This discussion leads on to one final principle, namely that of nomenclatural transfers. Poeppig's original name for the giant water-lily of the Amazon was *Euryale amazonica,* the specific epithet of which has priority. But the species is now considered to belong to the genus *Victoria.* The ICBN rules that Poeppig's specific epithet must have priority since it is the oldest, but that it must be transferred to the correct genus to give the right name for the species, that is *Victoria amazonica.* The new combination was first published by Sowerby in 1850 and requires a new authority citation as follows: *Victoria amazonica* (Poeppig) Sowerby, which includes the authority for the specific epithet and the authority for the new combination. Similar rules govern change of rank or transfers within a species.

These rules for the scientific naming of plants are crucial to the work of all scientists interested in plants and especially so to the plant geographer. Not only must he safeguard the details of his plant collections and records but he must be certain that the plants collected and recorded are correctly named and classified. If there is a flaw in either of these processes, the work of the plant geographer will rest on unsure foundations. He is above all dependent on the work of the taxonomist, who identifies, classifies and names the plants, the distributions of which form the core of the subject 'plant geography'. Advances in taxonomic method are therefore one of the major stimuli to new research and theories in plant geography.

Recent developments in taxonomy

In his Presidential Address to the Systematics Association delivered in December 1973, Professor Heywood described systematics as 'the stone of Sisyphus' – the taxonomic rock, he argued, was growing in size and was becoming increasingly difficult to push up the hill! Since around 1960 taxonomy has experienced what has been referred to as a 'revolution' and classical taxonomy has come under close scrutiny, with writers like R. R. Sokal declaiming that the need for descriptive taxonomy has passed, an argument effectively rebutted by Professor Heywood in his address. Yet recent developments in taxonomy have profound implications for the work of systematic botanists and for all those disciplines that rely on their work, including plant geography.

Although it is not relevant, or feasible, to survey the whole field of recent taxonomic activity in this book, it is important to highlight four main areas of advance, namely: the application of high-speed electronic computers in the development of quantitative methods of classification (numerical taxonomy or taxometrics); the employment of electronic data processing (EDP) for the

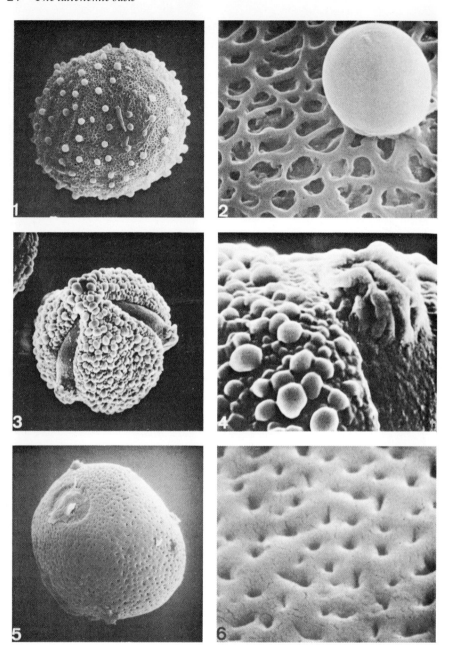

Figure 2.5 Scanning electron micrographs of the pollen morphology of four species of *Bauhinia* (Caesalpiniaceae) (*after* Larsen 1975, 119; reproduced by kind permission of Mrs Supee S. Larsen and the Editors of *Grana*).

   1, *B. hirsuta,* ×500. 2, *B. acuminata,* detail of sexine surface, ×6000. 3, *B. bassacensis,* ×1300. 4, *B. bassacensis,* detail of sexine surface near aperture, ×5000. 5, *B. bidentata,* ×1200. 6, *B. bidentata,* detail of sexine surface, ×11 000.

recording, storage and retrieval of taxonomic information; the use of scanning electron microscopy; and the growth in biochemical systematics. The latter two developments are different in kind from the first two. Scanning electron microscopy and biochemical systematics both produce new taxonomic data whereas the application of electronic computers provides new methods of handling the data available.

The use of the scanning electron microscope in taxonomic research became possible around 1965, following the previous application of the transmission electron microscope (TEM) since 1952. Scanning electron microscopy (SEM) permits the detailed observation – well beyond that possible with the light microscope – and photography of morphological features at high magnification and with great depth of focus, producing quasi-three-dimensional photomicrographs that can be readily interpreted. Consequently, SEM has revealed a whole new range of 'micro-characters' in the surface features of, for example, **pollen grains** (see Fig. 2.5), seeds and leaves, which may be of great taxonomic value. In a recent study of the pollen taxonomy of the beech genus (*Fagus*) and the southern beech genus (*Nothofagus*), Hanks and Fairbrothers (1976) found that the observation of the detailed surface sculpturing made possible by SEM enabled them to identify species within the *Fagus* pollen type where the data came from a single country or land mass. However, due to the range of sculpturing variation, this was not possible when all the *Fagus* taxa were taken together. In the case of *Nothofagus,* they went further and, using the pollen data obtained, produced what is probably the first computerised diagnostic key based on SEM information. There is little doubt that SEM will prove an important tool in helping to unravel some phytogeographical problems, especially the taxonomic and geographical distribution of pollen types (see pp. 84–5).

Similarly, within the last twenty years or so, biochemical systematics has proved a formidable new frontier, although the value of chemical characteristics in plants has long been recognised by taxonomists. The revolution has been primarily brought about by the development of techniques such as chromatography and electrophoresis which enable the rapid identification of large numbers of organic compounds and which may, in some cases, be used on herbarium material many years old as well as on fresh material. This new range of data is not only of high taxonomic value but also throws light on the common ancestry of plant taxa and is proving important in understanding the evolutionary pathways by which groups have arisen. Moreover, the study of phytochemical differences is increasingly important in plant geography, both in refining the specification of taxa whose distribution and evolutionary history are being analysed and also, in its own right, as a vital line of enquiry when tracing migratory routes and evolutionary sequences (see Fig. 2.6), assessing the ability of plants to affect potential competitors allelochemically, and considering the relationships between disjunct populations (see pp. 91–108). Bell and Evans (1978), for example, have recently invoked biochemical evidence to indicate a possible former link between the **floras** of Australia and the Mascarene Islands.

Studying the seeds of the legume genus, *Acacia,* which is widely represented in America, Africa, Asia, Australia and many oceanic islands, they discovered that a single characteristic non–protein amino–acid pattern characterised the seeds of some 67 species from Australia. Elsewhere, this pattern only reappeared in one non–Australian species, *A. heterophylla,* a native of the Mascarene Islands.

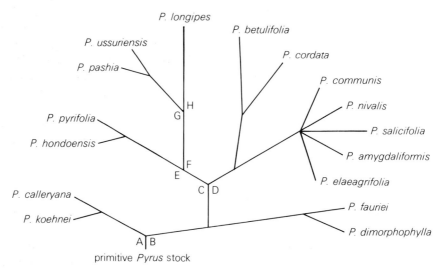

Figure 2.6   A possible system of branching sequences in the evolution of the pear genus, *Pyrus,* based largely on the occurrence of phenolics (*after* Challice & Westwood 1973, 145; © Linnean Society of London). A, retention of C_6-C_1 phenolic acid esters of calleryanin from original primitive *Pyrus* stock; B, loss of above; C, retention of flavone glycosides from original primitive *Pyrus* stock and gain of ability to synthesize flavone FS; D, loss of flavone glycosides; E, continuation of line (C); F, gain of ability to O-glucosylate flavones at the 4'-hydroxyl; G, retention of luteolin 7-rhamnosylglucoside; H, loss of above.

Both SEM and biochemical systematics have therefore proved rich sources of new taxonomic information that is of vital importance to plant geography also. But the boom in taxonomic data poses its own problems and is, in part, the reason for the increased application of high-speed electronic computers in both the development of quantitative methods of classification and the recording, storage and retrieval of taxonomic information. Taxometrics is concerned with the numerical evaluation of similarity between given organisms and their ordering into higher ranking taxa on the basis of these similarities, the computer doing the work without bias and at high speed. In such studies, the entities of lowest rank are known as Operational Taxonomic Units (OTUs), for which a minimum of between fifty and a hundred characters must be chosen to produce a satisfactory classification. For example, in a recent study of the *Bulbostylis/ Fimbristylis* (Cyperaceae) complex in Nigeria, Hall, Morton and Hooper (1976) used 55 characters (ranging from habit, through leaf and inflorescence characters, to the colour of the achene) with the interrelationships between the

OTUs being assessed by the use of principal components analysis.* Such studies can be important for plant geographers in that they may provide a significant refinement in the taxonomic base of the subject and they certainly facilitate the handling of the newer classes of data provided by the biochemists and molecular biologists.

Similarly, the more widespread application of electronic data processing for the recording, storage and retrieval of taxonomic information may also prove of great value to plant geography (Soper & Perring 1967), making a larger range of data more readily available. Yet the use of such data processing in museums and herbaria poses many problems, scientific, practical and economic, and as Professor Heywood pointed out in his Presidential Address to the Systematics Association referred to above, 'there have been very few electronic data processing projects so far where the data could not be produced as effectively and more cheaply by conventional means and retrieved by manual sorting'. The South African National Herbarium in Pretoria, however, is one major herbarium which extensively employs electronic data processing.

Advanced systems of recording

In contrast to the above experience, the usefulness of automatic data processing in the recording and storing of site information and in the preparation of distribution maps is now generally recognised. If it is important to safeguard and to refine the taxonomic basis of plant geography, it is no less vital to systematise and control the geographical records of the sites of discovery. This need not mean, of course, intensive and rigorous data collection on a grid system and a computer-mapped flora. Indeed for most parts of the world such a goal remains well out of reach. There is neither the prerequisite taxonomic knowledge nor the finance, and in the main most countries lack the necessary large band of enthusiastic and reliable amateur recorders. But this is no excuse for slapdash and inadequate recording and mapping. The fullest possible geographical data should be given on every herbarium sheet and with every published record. Often a coarser, less ambitious, but still systematic, method of recording – say by region or administrative district or, for example, by units like the Watsonian vice-county system discussed in Chapter 3 – may prove rewarding and an encouragement to more organised plant recording.

However, in countries and areas with a relatively small total flora and a long history of survey, such as the British Isles, the application of more advanced systems has already proved of value. In the British Isles, the reliable recording of vascular plants goes back to at least the end of the sixteenth century, and by 1629 Thomas Johnson had published the first 'local' flora with his *Iter Plantarum . . . in Agrum Cantianum,* which describes a botanical journey made through

*For thorough mathematical accounts of numerical methods used in biological taxonomy, see Sokal and Sneath (1963), Jardine and Sibson (1971) and Sneath and Sokal (1973).

Kent earlier that year (Gilmour 1972). The first approximation to a 'county flora' was John Ray's *Catalogus Plantarum circa Cantabrigiam nascentium* published in 1660, which was the result of nine years of labour and which marked the start of a great tradition of plant recording (see Raven 1942, Ch. 4). This tradition was furthered by the development of Watson's vice-county system (see pp. 36–8), originally proposed in 1852, which reigned supreme until the herculean efforts of the Botanical Society of the British Isles (BSBI) brought about its demise with the first advanced system of recording. Between 1954 and 1960, the BSBI co-ordinated what was probably the largest single exercise in recording ever undertaken by field botanists. The ten-kilometre square of the national grid formed the basic unit of survey and each square was assigned to a particular recorder or came under the scrutiny of a group effort. In all, 3500 10 × 10 kilometre squares were surveyed, yielding an average of over 400 species records per square, the records being marked on specially prepared data cards. But the real basis for this new and intensive method of survey was the mechanised sorting of punched index cards which made possible the manipulation of great amounts of information.

As might be expected this form of survey is being increasingly employed and further refined. In Britain it is being used to map a wide range of plant groups, such as lichens and mosses, as well as animals, and it has become the basis for most of the newer county floras. In some of these, such as Edees' *Flora of Staffordshire* published in 1972, the ten-kilometre squares are now divided into blocks of four one-kilometre squares termed tetrads (Fig. 2.7), an example of increasing refinement in relation to the scale of operation. Packham *et al.* (1979) intend to go one stage further in their proposed new Flora of the Shropshire region. This will cover a rectangular block of ten-kilometre squares, but records for species in three 'frequency classes' will be collected in different ways. Common (A) species will be recorded only at the ten-kilometre square level. Intermediate (B) species, however, will be mapped in 2 × 2 kilometre tetrads, but with records localised to the one-kilometre square, and rare (C) species will be given their six-figure Ordnance Survey grid references. Special field cards have been prepared for recording each of the three categories. The study of Warwickshire by Cadbury *et al.* (1971) is a good example of a local computer-mapped flora. In Belgium, Luxembourg and France, similar methods have become in vogue, but using an eight-kilometre square grid (see Fig. 3.7, Ch. 3).

This discussion of methods of recording and survey leads on directly to the next chapter, which considers the different techniques for mapping plant distributions. The more closely the method of survey is integrated with the cartographic techniques to be employed, the better will be the result, especially where automatic data processing is employed. It cannot be emphasised too fully that a satisfactory system of recording is the only sound basis for mapping plant distributions and thus for all work in plant geography.

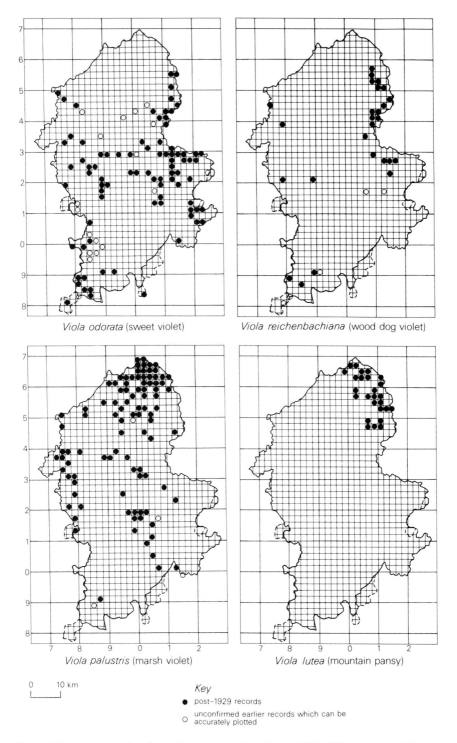

Viola odorata (sweet violet)

Viola reichenbachiana (wood dog violet)

Viola palustris (marsh violet)

Viola lutea (mountain pansy)

0   10 km

Key

● post–1929 records

○ unconfirmed earlier records which can be accurately plotted

Figure 2.7 An example of tetrad mapping (*after* Edees 1972, 203; reproduced by kind permission of the author).

Further reading

Useful introductory texts are marked with an asterisk.

Plant hunting
Kingdon Ward, F. 1937. *Plant hunter's paradise*. London: Jonathan Cape. (A very readable and enjoyable example of the art of plant hunting by one of its past masters.)
Lemmon, K. 1968. *The golden age of plant hunters*. London: Phoenix House.
*Whittle, T. 1970. *The plant hunters*. London: Heinemann.

Some more modern approaches
Cadbury, D. A., J. G. Hawkes and R. C. Readett 1971. *A computer-mapped flora. A study of the county of Warwickshire*. London & New York: Academic Press.
Packham, J. R., P. H. Oswald, F. H. Perring, C. A. Sinker and I. C. Trueman 1979. Preparing a new Flora of the Shropshire region using a federal system of recording. *Watsonia* **12**, 239–47.

History of a herbarium
Kalkman, C. and P. Smit (eds) 1979. Rijksherbarium 1829–1979. *Blumea* **25**, 1–140.

Botanical nomenclature and taxonomy
*Heywood, V. H. 1976. *Plant taxonomy*, 2nd edn. London: Edward Arnold.
Heywood, V. H. (ed.) 1968. *Modern methods in plant taxonomy*. London & New York: Academic Press.
Heywood, V. H. 1974. Systematics – the stone of Sisyphus. *Biol. J. Linn. Soc.* **6**, 169–78.
Jardine, N. and R. Sibson 1971. *Mathematical taxonomy*. New York: John Wiley.
*Jeffrey, C. 1968. *An introduction to plant taxonomy*. London: J. & A. Churchill.
*Jeffrey, C. 1977. *Biological nomenclature*, 2nd edn. London: Edward Arnold.
*Jones, S. B. and A. E. Luchsinger 1979. *Plant systematics*. New York: McGraw-Hill.
Juniper, B. E., A. J. Gilchrist, G. C. Cox and P. R. Williams 1970. *Techniques for plant electron microscopy*. Oxford: Blackwell Scientific.
*Radford, A. E., W. C. Dickison, J. R. Massey and C. R. Bell 1974. *Vascular plant systematics*. New York: Harper & Row. (Chs 1–4 and 10.)
Smith, P. M. 1976. *The chemotaxonomy of plants*. London: Edward Arnold.
Sneath, P. H. A. and R. R. Sokal 1973. *Numerical taxonomy*. San Francisco: W. H. Freeman.
Stearn, W. T. 1973. *Botanical Latin*, 2nd edn. Newton Abbot: David & Charles.

3 Plant maps

Because the plant geographer is essentially concerned with the spatial distribution of plants, the first step in analysing the basic information provided by collecting and recording is the preparation of plant distribution maps. These depict cartographically the recorded range of the plants in question and represent initial generalisations of the original information. On only very few maps is it possible to include every individual point at which a plant occurs, and therefore from the outset the plant geographer must exercise his judgement on the way he organises his material for cartographic purposes. Nearly all plant maps will involve the simplification and regrouping of the information to be shown, and carelessness at this stage will not only produce unhelpful and misleading maps but could also jeopardise the validity of theories put forward to explain the distributions concerned. Unfortunately, many plant geographers pay insufficient attention to fundamental cartographic principles, and published maps should be viewed as critically as their accompanying texts are read.

Three basic principles

When either preparing a plant distribution map or making an appreciation of a finished product, it is essential to take into account three basic cartographic principles. These principles should govern the way in which a plant geographer approaches the cartographic generalisation of his information. In most cases this process will include the simplification of that information, its reorganisation and classification and its graphic summary on a map. Although in practice it would be artificial to isolate these three principles from each other, for the sake of convenience they are treated separately below.

Principle 1. A map must always reflect the quality of the information on which it is based. It is especially important in cartography that the final map does not give an impression of accuracy greater than the source material warrants. Psychologically a map is a very definitive product and erroneous or doubtful plant records can soon change into accepted and unchallengeable 'facts' when mapped uncritically. In plant maps it is essential to differentiate *on the map* records from different dates and periods of taxonomic knowledge, and records of doubtful identification or geographical location. It may well be necessary to distinguish records based on herbarium material from those derived from less reliable sources, and identifications checked by accepted authorities from those made by amateur recorders. The dot maps of the distribution of *Pedicularis sceptrum-carolinum* in southern Sweden (Nilsson & Gustafsson 1976) and of ash

(*Fraxinus excelsior*) in Denmark (Ødum 1968) shown in Figure 3.1 are models of the care needed in this respect to produce a good map. Moreover, the whole map should reflect the quality of the information under consideration. The mapping technique used will itself partly depend on the nature of the data and it

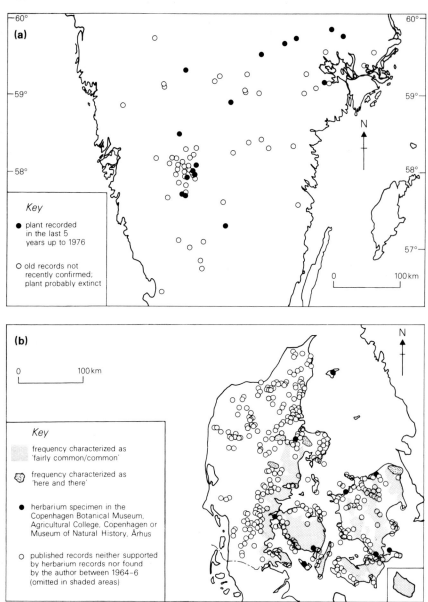

Figure 3.1   Dot maps of the distribution of (a) *Pedicularis sceptrum-carolinum* in southern Sweden (*after* Nilsson & Gustafsson 1976, Fig. 4) and (b) ash (*Fraxinus excelsior*) in Denmark (*after* Ødum 1968, Fig. 56). Note the care with which the plant records have been edited.

is always important to distinguish between records for regions that have been thoroughly researched and surveyed and those for areas not so intensively studied. The 'presence-and-absence' value of a record is governed by the intensiveness of the survey on which it is based. A single record from a poorly known region tells us little and should not be made to imply too much.

Principle 2. The scale of the map is the main factor determining the nature and degree of generalisation that should be employed. All maps are reductions, including the thematic maps of the plant geographer, and it is the degree of this reduction which more than anything else determines the level of generalisation that is necessary and acceptable. In most cases, the smaller the scale, the greater will be the degree of generalisation required. The plant geographer should, therefore, choose scales that are in keeping with the reliability and intensity of his information and that are best suited to the basic aim of the map being designed. Moreover, as in any well constructed map, the scale being used should *always* be indicated, whether by means of a verbal statement, bar scale, area scale or representative fraction. Without this information it is impossible to interpret a plant map satisfactorily, for, after all, plant geography is primarily concerned with the geographical distribution and areal relationships of plants. Unfortunately, many published plant maps lack this marginal note of the scale, and in their preparation little attention has been paid either to the choice of scale used or to the amount of detail that may be portrayed successfully at the scale chosen.

Principle 3. The overall design of the map should be closely related to the basic purpose of the map. The aims behind plant maps are not always the same. On many, it is true, the idea is simply to show, in very general terms, the world, regional or local distribution of a given taxon or group of taxa (e.g. the map of the distribution of *Schuurmansia* spp. in Figure 3.3). But on some plant maps, the aim is much more specific. The author may wish to demonstrate the possible correlation between a particular distribution and an important geological, climatic or historical factor or he may try to highlight the relationships between certain plant populations or taxa. Moreover, it is possible to draw maps that show suggested migratory or evolutionary pathways (Fig. 3.2) or that plot the changing pattern of a distribution through time. Yet, whatever the aim may be, it is essential that the purpose is clearly defined in the cartographer's mind *before* he embarks on the map. The aim of the map will govern the cartographic techniques used, its scale, the nature of the projection where needed (see p. 41) and the general style of the map. The aim itself will be partly determined by the type and quality of the information available, and perhaps the most crucial point of all is the need for *consistency* of generalisation throughout the whole process of map making. In plant maps the plant geographer must strive for a close harmony between the quality of his basic information, the aim of the map and the chosen scale at which it is to be drawn. Only where these are in tune will a plant map have any chance of success.

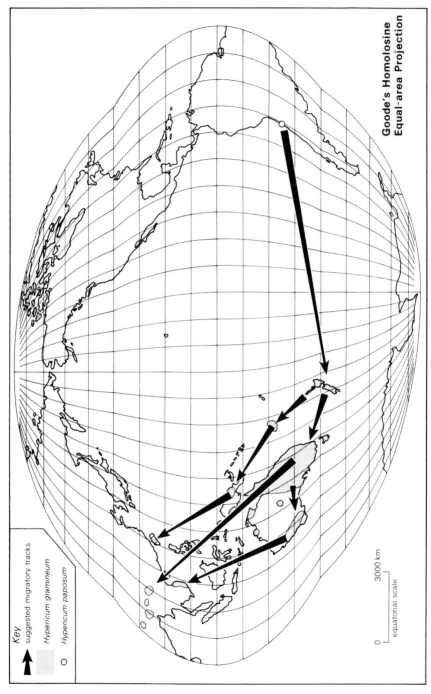

Key
suggested migratory tracks
Hypericum gramineum
○ *Hypericum paposum*

0   3000 km
equatorial scale

Goode's Homolosine
Equal-area Projection

Figure 3.2 The suggested migratory tracts of *Hypericum gramineum*, based on trends in morphology. The point of origin indicated in northern Chile is the location of *H. paposum* (after Robson 1972, Fig. 3). Note the choice of projection and the presence of a bar scale.

Methods of mapping plant distributions

Historically, plant geographers have used a number of techniques for making plant maps. Although these are treated separately below, they are often employed in combination, and where such multiple techniques have proved important they will be discussed under the appropriate headings.

Mapping plant distributions by outline areas. These are the simplest, least accurate and most general of all plant maps, although they are widely used, especially in taxonomic works. The aim of the technique is to show by means of continuous lines the *boundaries* of a given plant distribution (Fig. 3.3). Only rarely is the nature of the distribution within these geographical boundaries indicated. The topography of the distribution, its varying density and its degree of disjunction are usually ignored. In many cases an arbitrary decision is taken to distinguish outlying areas or localities from the main compact area, and sometimes selected characteristics of a distribution, such as known extinctions, are shown within or outside the present boundary. Occasionally the outlined areas are shaded or blacked-in (e.g. Fig 3.8) and for greater accuracy broken lines and question marks can be used to indicate portions of the boundary that are unknown or uncertain, or areas for which the records are doubtful (as in Figure 3.3).

Paradoxically, to be in any way acceptable this technique depends on a fairly thorough knowledge of a distribution for the final map is presented as an ultimate generalisation of the geographical limits of a taxon. Too frequently,

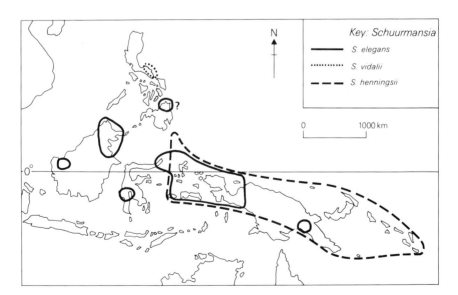

Figure 3.3 An outline distribution map of *Schuurmansia* spp. in the Indo-Pacific area. Outline areas are only slightly generalised from cited localities (*after* Kanis 1968, Fig. 8).

and wrongly, the method is used to link isolated and rare records from little researched areas, records which need to be far more carefully distinguished on the map (Principle 1). Another problem is encountered on maps showing the distribution of more than one taxon, for the proliferation of too many lines and types of line soon leads to confusion and the maps become unreadable.

For recording general distributions, outline plant maps are really only appropriate on small-scale continental or world maps where a more thorough and accurate method is neither possible nor necessary (Principles 2 & 3). They are also useful, however, in generalising cartographically more detailed maps when it is wished to investigate the relationships beween different parts of a distribution or to illustrate the correlation between the boundary of a distribution and a certain environmental measure, such as a given **isotherm** or **isohyet**. It should always be remembered that an outline plant map is little more than a map of the known geographical limits of a taxon.

Mapping plant distributions by administrative units or recording districts. At the national level, mapping plant distributions by administrative units or recording districts has proved a valuable method, especially in countries where there has been a long history of recording and where plant distributions are reasonably well known, though still under-recorded for more detailed mapping purposes (Principle 2). It has the advantage of closely linking the methods of recording and mapping and provides a systematic basis for further research. Moreover, the mapping units may have great permanence historically, and this facilitates the use of past records, although with such records there are invariably difficulties of identification. On the map each unit for which the plant has been recorded is differentiated, usually by means of shading.

Such maps, however, suffer from a number of serious limitations. Most do not distinguish between units in which the plant has been recorded only once and units in which it is common. There is consequently little indication of the topography or intensity of a distribution. Moreover, the mapping units used are rarely of equal area and seldom coincide with natural physiographic boundaries. As a result the final map frequently tells us nothing about the real ecological and geographical distribution of a plant, although it does summarise cartographically the known records of its occurrence. Indeed, in certain circumstances, the map can be positively misleading about the nature of a distribution. This is particularly so for coastal distributions which are mapped in units that extend far inland.

In Britain the tradition (mentioned in Ch. 2, p. 28) of compiling county floras led on naturally to this form of mapping, a tendency that was furthered by the development of the Watsonian vice-county system. This was a system, originally proposed in 1852 by H. C. Watson, in which the county provided the basic recording unit, but which also recognised that there was too great an areal discrepancy between the larger and smaller counties. The bigger counties were therefore divided into a number of smaller units and some small counties linked with neighbours in an attempt to create a more even base for recording. Each of

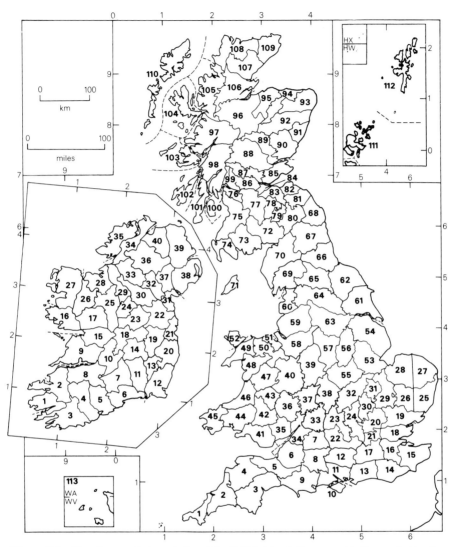

Figure 3.4 The vice-county boundaries of the British Isles (*after* a map kindly supplied by the Biological Records Centre of the Institute of Terrestrial Ecology).

the new vice-counties was given a name and a number (Fig. 3.4). For example, Kent was divided into two vice-counties (V.-c. 15 East Kent and V.-c. 16 West Kent), and V.-c. 55 was Leicestershire (with Rutland). In all, excluding Ireland, which was later treated in a similar fashion by Praeger (1896), Watson defined 112 vice-counties, not including the Channel Islands, which are designated '113' or 'S' (for Sarnia) in some publications. Until the late 1950s botanists took great pleasure in discovering a new species for a given vice-county, and many specialist journals carried regular lists of new vice-county records.

Although this system possessed a great deal of continuity historically, problems still arose over defining the precise boundaries of different vice-counties. Of course this difficulty was much exacerbated by changes in the administrative boundaries of the counties since Watson's time, and in 1947–8 the Systematics Association formed a Subcommittee on Maps and Censuses, the job of which was to fix accurately the vice-county boundaries of Great Britain. The results gave rise to a definitive account of the boundaries published by Dandy in 1969. This was accompanied by two Ordnance Survey maps at a scale of 1 : 625 000 on which the agreed lines were marked. In 1949 the Ordnance Survey of Ireland published a similar map showing the boundaries of the Irish vice-counties, as defined by Praeger.

Systems such as these may soon be of great value in recording and mapping the floras of developing countries. Although the technique has its limitations, it could be refined further and there is scope in plant geography for a wider use of true choropleth maps, in which some statistical measure is applied to the degree of presence of a taxon within the enumeration districts. A simple choropleth map is '. . . a spatially arranged presentation of statistics that are tied to enumeration districts on the ground' (Robinson *et al.* 1978, 244). It may well be possible to devise indices of plant presence (and absence) which would provide such a statistical basis. The analysis of population size within each district could be one approach. The production of simple choropleth maps might, in turn, lead to the extension of dasymetric and isarithmic mapping techniques into plant geography. Basically, these are all methods of representing quantitative data by either line symbols (isarithmic maps) or area symbols (the simple choropleth and the dasymetric map). The techniques are described in full by Robinson *et al.* (1978, 217–57).

Mapping plant distributions by dots. Even though the utopian situation of one dot on a map actually representing one plant record in the field is rarely achieved, the dot map remains a most important and accurate way of mapping plant distributions. It is, however, much abused. The main reason for this abuse lies in the fact that virtually no attention is paid to either the value or the size of the dots themselves. On many plant maps, a single dot could mean anything from one record to a thousand, and the arbitrary way in which records are turned into dots is seldom clear or explained. Yet, if the visual impression of a dot is not to be misleading, its unit value and size must be chosen carefully and stated *on the map*. This unit value and size will depend on the nature and quality of the basic information being used (Principle 1) and on the final scale of the map (Principle 2).

Dot size is primarily a matter of good or bad cartography, although dots of an ill chosen size can certainly give a false impression of the nature of a distribution. If the dots are too small, the distribution will appear sparse and patterns will not emerge. If they are too large, they will coalesce and the distribution will appear excessively dense. Ross Mackay (1949) has developed a useful nomograph to assist in determining the most satisfactory dot size and density, and many plant

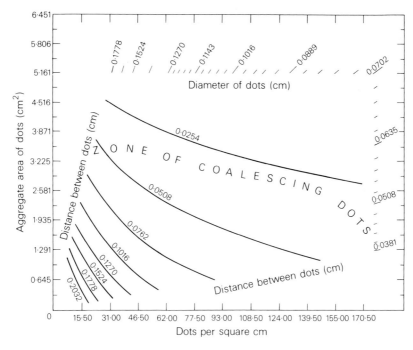

Figure 3.5 A nomograph to show the relationship between dot size and dot density. This should be used to determine the desirable dot size and unit value (*after* Ross Mackay 1949; taken from Robinson, A., R. Sale & J. Morrison 1978. *Elements of cartography*, 4th edn. Reprinted by permission of the Executive Director of the American Congress on Surveying and Mapping and of John Wiley & Sons, Ltd).

geographers would do well to use such an aid in the preparation of their dot maps. This nomograph (Fig. 3.5) requires a knowledge of the sizes of dots that can be made with various types of pen. By varying the relationship between dot diameter and unit value, the graph can be employed to find a compromise between the two that will best present the characteristics of the distribution.

The value of the dot is a matter of particular concern, for it relates to the manner in which the basic information of the plant geographer is organised and generalised to produce plant maps. Yet there are few plant maps indeed on which the value of the dot is stated. An exception to this is found in the work of Selander (1950) on the floristic phytogeography of south-western Lule Lappmark. Figure 3.6 reproduces a map compiled by Selander to show the distribution of *Cassiope tetragona* at its southern Swedish boundary. On this map three sizes of dot are used to indicate, albeit subjectively, the degree of presence of the plant at each recorded locality. The largest dot signifies the *Cassiope tetragona* heath community, in which the plant is obviously common. An intermediate dot indicates sites at which it is fairly plentiful, and the smallest dot size refers to localities with single tufts only. It is also worth noting the care with which Selander edits his records (Principle 1). Filled dots, in all three size categories, represent sites actually visited by the author. Open circles are stations taken from the literature or based on herbarium records that have not been confirmed by the author. Moreover, the map carries an indication of its scale and is well drawn. It is, without doubt, a satisfactory plant map, soundly based on the three basic cartographic principles enunciated at the beginning of this chapter.

Of course, although the dot is the simplest of all point symbols, it is not the only one, and on maps showing the distribution of more than one taxon it is often replaced or accompanied by a wide range of alternative point symbols, such as crosses, squares and triangles (e.g. Fig. 8.4, Ch. 8). Unfortunately, many of these maps are too confusing to read easily, and as a general rule maps with more than two or three different symbols on them tend to defeat their own object and the information would be better conveyed in the text. Moreover, point symbols are frequently used in combination with a wide range of other techniques (e.g. Fig. 3.1b, Fig. 1.1), such as outline plant areas and recording districts, but, above all, in conjunction with grids.

Mapping plant distributions by grids. One of the major cartographic developments of the 1914–18 war was the use of grids for defining the positions of points. Strictly speaking a grid is a graphic representation of a plane rectangular coordinate system which is used for defining the positions of trigonometrical points on a projection. Because the formulae of plane geometry are much

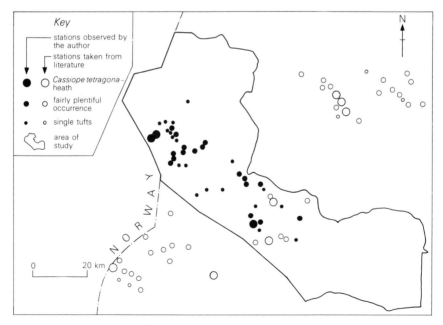

Figure 3.6   The distribution of *Cassiope tetragona* at its southern Swedish boundary (southwestern Lule Lappmark) (*after* Selander 1950, Fig. 19). Note how a series of dot sizes is used to indicate varying degrees of presence.

simpler than those of spherical geometry, many countries soon devised systems of plane rectangular coordinates, such as the grid system employed by the Ordnance Survey of Great Britain and the UTM (Universal Transverse Mercator) or UPS (Universal Polar Stereographic) military grids adopted in the United States. Not surprisingly, plant geographers were quick to see that such grids could provide a very accurate and flexible method for both the recording (see Ch. 2, p. 28) and the mapping of plant distributions.

As pointed out by Seddon (1971) a grid system has many advantages. In the first place, the technique of recording can be closely integrated with the method of mapping, although this is not essential. Secondly, records from adjacent areas can be amalgamated and turned into larger unit squares. Indeed the individual unit of the grid can be varied to show whatever degree of detail is best suited to the scale of the finished or published map (Principle 2), the only limitation being the original mesh size of the initial survey – it is obviously not possible to subdivide further without a more detailed survey based on a finer mesh. Moreover, the system lends itself easily to the needs of electronic data processing, and the basic information is readily re-sorted automatically with the maps produced by computer printout. It is therefore a technique increasingly used in the more advanced countries.

A recent example of such a grid-mapped flora is that of the *Atlas de la Flore Belge et Luxembourgeoise* (Rompaey & Delvosalle 1972). In this *Atlas*, 1626 taxa of flowering plants and ferns are mapped for 2970 16 km² squares based on the *'Carte d'État-major de Belgique au 1 : 40 000'* (now 1 : 50 000). The essential field work for this project was carried out between 1930 and 1970 by collaborators of the Institut Floristique Belge (IFB) and recording took place in 1 km² squares derived from the 16 km² squares of the main grid. Figure 3.7 reproduces Map 1248 from this *Atlas* and shows the distribution in Belgium and Luxembourg of the bird's nest orchid (*Neottia nidus-avis*).

In addition to simply shading-in the square for which a plant is recorded, a wide range of symbols may be used within squares for distinguishing records from different dates (as in Fig. 1.1) or records of varying validity or for indicating the distribution of more than one species on the same map. At the more detailed level, a grid map is really a sophisticated and systematic dot map. In such instances, the comments made earlier (p. 38) concerning the importance of dot size and value are just as apposite, and one significant way of refining the grid map would be to provide some indication, quantitative if possible, of the degree of presence and absence of a given taxon within each square.

Plant maps and map projections

Like any other map, the success of a plant map will depend to a considerable extent on the choice of a suitable map projection. This is especially so in the grid maps just discussed, where the grid must be actually based upon and closely related to a projection, but it is also true for nearly all types of plant map,

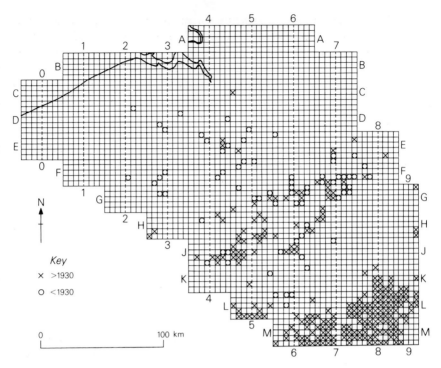

Figure 3.7 Map 1248 from the *Atlas de la Flore Belge et Luxembourgeoise* showing the distribution of the bird's nest orchid (*Neottia nidus-avis*) (*after* Rompaey & Delvosalle 1972; reproduced by kind permission of The Head of Department, Spermatophytes-Pteridophytes, Jardin botanique national de Belgique. *Note:* a 2nd edn (1979) of the *Atlas* has now been published in which map 1248 has been revised). An example from a grid-mapped flora.

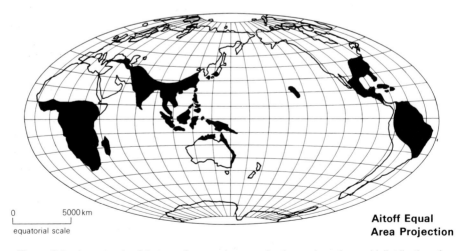

Figure 3.8 An example of the use of an equal-area projection to show the world distribution of a taxon, in this instance the Pantropical genus *Diospyros*. The projection is Aitoff's Equal-area projection and the bar scale gives distances as measured at the Equator (*after* Thorne 1972, Fig. 19).

particularly those showing distributions at the continental, hemispheric or world scales. The choice of projection will depend on a number of factors, including the main purpose of the map, the shape of the area to be represented and an appreciation of those map projection qualities that are most suited to the basic nature of the map. Although it is impossible to make a perfect map of any portion of the globe, it is feasible to preserve certain qualities in a projection, such as bearing, scale, shape (orthomorphism) or area (Steers 1970). However, since one is unable to preserve all these qualities in a single projection, the cartographer must choose which of these qualities it is most essential to maintain in his particular map and then select an appropriate projection.

Because plant geographers are primarily concerned with distributions, it is usually logical to employ an equal-area projection (e.g. Figs 3.2, 7.1, 7.2, 7.6), for this will facilitate regional comparisons of area. At the world scale, both Mollweide's Homolographic and Hammer's (or Aitoff's) Equal-area projections (Fig. 3.8) are suitable and indeed the former was used by Good for the majority of the maps in his *The geography of the flowering plants* (4th edn 1974, pp. 19–20). A third possibility is Sanson–Flamsteed's (Sinusoidal) Equal-area projection, but the shape of the land masses is not as well preserved as on Mollweide. If the purpose of the map allows, all three projections may be greatly improved by interruption over the oceans (Figs 3.9, 7.9).

However, if a rectangular world map is preferred to the ellipses of the above, it is probably best provided by a general projection such as Gall's (stereographic) projection, which is neither equal-area nor orthomorphic, but which exhibits far less exaggeration in polar areas than the ever popular, but much misused, Mercator's projection. Because of the great areal exaggeration in high latitudes, Mercator is unsuitable for showing land masses and will give a totally false impression of the relative areas occupied by plants in, for example, the boreal and equatorial regions (Roblin 1969). It is properly employed for sea and air navigation and for any other purpose (e.g. mapping ocean currents) in which correct direction is important.

For mapping distributions in polar regions, Lambert's Polar Zenithal Equal-area projection is possibly the most useful, although for the tropics any equal-area or general world projection which shows the hot areas well would be appropriate. Africa, however, is a special case in so far as it lies astride the Equator, and it is best represented on either Lambert's Equatorial Zenithal Equal-area projection or Sanson–Flamsteed's projection. But there can be no hard and fast rules. Virtually every plant map or series of plant maps will necessitate a distinctive choice of projection, especially if centred on a particular continental area.

There is probably a great deal of room in plant geography for a wide range of experiment in the use of map projections. It would be fascinating, for example, to plot a given distribution on a number of well known and unusual projections and then to compare the results. The differences, both major and minor, would become immediately apparent. But whatever projection is finally chosen, like the scale of the map, it should always be indicated on the finished product.

Moreover, it cannot be emphasised too strongly that a plant geographer would always do well to consult a globe, which is the working model that will mislead him the least.

Some recent developments

It is perhaps true to say that research into the improvement of plant mapping techniques has been neglected. Yet in recent years this has been remedied to some extent, especially with the development of computational methods (Cadbury *et al.* 1971, Soper & Perring 1967). In addition, new ideas have been put forward for increased efficiency in the manual mapping of locality records.

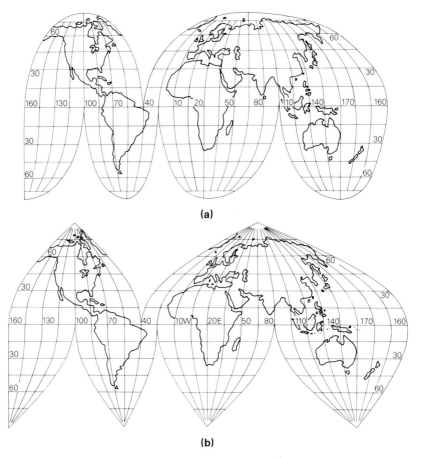

(a)

(b)

Figure 3.9 Two interrupted equal-area projections: (a) interrupted Mollweide projection (Goode's homolographic projection), with the breaks on the oceans and the equal-area property of Mollweide preserved; (b) interrupted Sanson–Flamsteed projection (reproduced by permission of Hodder & Stoughton Ltd from Steers, J. A. 1970. *An introduction to the study of map projections,* 15th edn).

One such scheme is that proposed by Werner (1977), which relies on the use of bi-coloured base maps printed in quantity. Geographical features that will be retained on the finished map are marked in black, with other features, in particular a grid and reference numbers, printed in pale blue. By using the grid, locality records are quickly and accurately entered on to the map. Yet, on publication, there is no need to retain this grid, for the blue lines are easily filtered out. Such simple suggestions can much improve both the speed and accuracy with which plant records are turned into plant maps.

Another area in which there have been welcome developments is in the compilation of indexes of published plant maps. These indexes not only facilitate research into plant distribution but also help to avoid the duplication of maps and guard against incomplete surveys of what is already known about a particular distribution. A good example is the *Index des cartes de répartition. Plantes vasculaires d'Afrique (1935–76)* (Lebrun & Stork 1977), which catalogues all the published maps for no less than 8830 African taxa. Because plant maps are printed in so diverse a range of publications, such compilations have become increasingly necessary and will become, like the plant maps themselves, an indispensable tool of the plant geographer.

Further reading

Useful introductory texts are marked with an asterisk.

*Keates, J. S. 1973. *Cartographic design and production*. London: Longman.
Perring, F. H. 1968. *Critical supplement to the atlas of the British flora*. London: Nelson.
Perring, F. H. and S. M. Walters 1962. *Atlas of the British flora*. London: Nelson.
*Roblin, H. S. 1969. *Map projections*. London: Edward Arnold.
Robinson, A., R. Sale and J. Morrison 1978. *Elements of cartography*, 4th edn. New York: John Wiley.
Seddon, B. 1971. *Introduction to biogeography*. London: Duckworth. (Chapter 1.)
Steers, J. A. 1970. *An introduction to the study of map projections*, 15th edn. London: University of London Press.
Werner, Y. L. 1977. Manual mapping of locality records – an efficient method. *J. Biogeog.* **4**, 51–3.

4   *Patterns of distribution*

The plant distribution maps discussed in Chapter 3 represent the first stage in the phytogeographical analysis of the basic information collected from the field and available in the herbarium and library. On these maps the geographical pattern of a distribution is summarised cartographically. The careful inspection of such maps will soon reveal that certain patterns tend to recur, although no two distributions are absolutely identical in all their aspects. The next stage in phytogeographical analysis is therefore an attempt to identify and classify these recurring patterns. In the past the recognition of types of pattern was basically intuitive, but in more recent years there has been an increasing application of computers to the analysis of geographical distribution in plants (e.g. Jardine 1972, Sneath & Sokal 1973, Sokal & Oden 1978). Both the traditional approach and the mathematical and computational approach are considered in this chapter.

The patterns of distribution recognised fall into two broad categories. In the first of these, the plant geographer is concerned to define the affinities between the distributions of different taxa. In the second, the interest derives from an analysis of the affinities between different geographical areas. The two methods are discussed separately.

Affinities in the distributions of taxa

Floristic elements. The traditional approach to the identification of affinities between the distributions of taxa involves the recognition of what are termed floristic elements, which are also known as floral elements or phytogeographic elements. The basic aim of the method is to group the taxa that comprise the flora of a given geographical area, be it an island, a country or a continent, into characteristic types or elements, defined according to one overriding criterion. Historically, a number of such criteria have been used and have given rise to five main types of floristic element (Polunin 1960).

Ecological elements are grouped according to their habitat preferences. For example, a distinction based on climatic affinities may be made between the oceanic and the continental elements within a flora. Likewise, a pedological division can be recognised between taxa characteristic of calcareous soils (the calcicole element) and those which avoid such substrates (the calcifuge element). *Historical elements,* in contrast, are defined with reference to the geological time period in which they became an integral part of the flora concerned, for example, in northern Europe, the Arctic-Tertiary element of evergreen and deciduous trees and the Boreal-Tertiary element, which included southerly members like the palms. *Migration elements* are arranged according to the routeways by which they spread into the region under consideration; whereas *genetic elements* are grouped according to their centres of origin (see Ch. 5). But the most important and widely used type of floristic element is the *geographical element,* or biogeographical element, a concept which in most circumstances also embraces some of the criteria already mentioned for the previous four types.

Geographical elements are defined by reference to the geographical range of taxa as mapped. In analysing the geographical distributions of the taxa that comprise the flora of a particular territory, it will be seen that the individual distribution maps for each taxon exhibit an areal bias. In very simple terms, for example, some taxa will be confined to the north of the territory, others to the south, east or west, and others again may possess a very general distribution throughout. By grouping together taxa whose distributions show similar, though of course rarely identical, areal bias, it is possible to define the major geographical elements within the territory. As early as 1873, Watson had recognised six such elements for the British flora. His 'British type' included species that were found throughout England, Wales and Scotland. He then distinguished 'English' and 'Scottish' types, as well as a 'Highland' type that comprised species restricted to mountain areas. Finally, he defined a 'Germanic' type, with species confined to east and south-east England, and an 'Atlantic' type, embracing a group of species found mainly in the west and south-west of Britain.

This analysis of the different areal patterns found *within* a small territory is the simplest form of the geographical element, and in consequence the least informative. Except for endemic taxa, all groups will have distributions which extend well beyond the boundaries of the territory under consideration, and a more complex analysis must take into account the total distribution of taxa, both within and outside the region actually under study.

Such is the case in Figure 4.1. The imaginary territory A at the centre of this theoretical example has a total flora of but six species of vascular plants, only one of which is confined in its distribution to the territory itself. Obviously, therefore, this represents the endemic element in the flora. The distributions of the other five species, although exhibiting areal bias within the territory, also extend beyond the boundaries of the territory into adjacent regions. In fact, for all five species, the main part of the distribution lies outside territory A and in a different area. Species *b*, for example, only just reaches territory A, the bulk of its distribution being to the north-east in territory B. This is therefore the territory B floristic element in the flora of territory A. The same argument applies to the distributions of species *c*, *d* and *e*, which constitute respectively the territory C, D and E floristic elements. The distribution of species *f*, however, is a little different in that it extends throughout the northern regions of this imaginary world and thus represents a general northern element in the flora of territory A. If there were also a species *g*, that was found in all areas, including territory A, this would be the cosmopolitan element in the flora.

From this theoretical example, we can see that more complex geographical elements are defined by reference to the location of the main part of their total

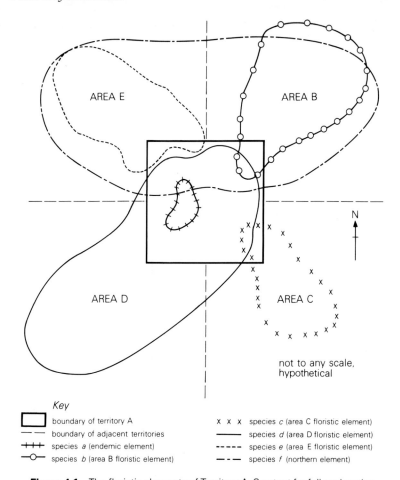

Figure 4.1   The floristic elements of Territory A. See text for full explanation.

distribution, regardless of their actual distribution within the territory under consideration. Obviously, in reality, the plant geographer is normally dealing with more than one taxon in each element and he is therefore looking for distributions that are broadly centred in the same region and whose boundaries extend in basically the same directions. Such taxa can then be grouped together as one geographical element, the element being named either after the core area of the distribution or after a particularly characteristic taxon or set of taxa.

In a recent study of the phytogeography of New Guinea (including both Papua New Guinea and Irian Jaya), van Balgooy (1976) recognised fifteen distribution types or elements within the flora, based on a detailed analysis of the total distribution of all genera represented by indigenous species (Table 4.1). For this study the genus proved the most suitable unit because the 9000 species are little known so far and because there is less difference of opinion among taxonomists as to the circumscription of genera. In fact, in most statistical

Table 4.1 A classification of the distribution types and the distribution type spectrum of New Guinea genera (*after* van Balgooy 1976, 10–11).

Distribution type	Typical representative	Number of genera with species in New Guinea	% Total flora
1. Cosmopolitan/pantropical	*Cyperus*	307	21·0
1a. Temperate wides	*Euphrasia*	39	2·7
2. Amphipacific tropical	*Batis*	30	2·0
2a. Northern temperate	*Rhododendron*	27	1·8
3. Old World generally	*Pittosporum*	294	20·1
4. Palaeotropical & Indo-Malesian, reaching Australia	*Balanophora*	89	6·1
4a. As above, but not in Australia	*Kopsia*	221	15·1
5. Strictly Malesian	*Dendromyza*	82	5·6
6. Subendemic genera, centred in New Guinea	*Archidendron*	97	6·6
6a. Strictly endemic	*Annesijoa*	98	6·7
7. Australian centred	*Hibbertia*	36	2·5
7a. Australian–New Guinea	*Eupomatia*	47	3·2
8. Pacific centred	*Ascarina*	24	1·6
8a. Australian–Pacific	*Quintinia*	46	3·1
9. Subantarctic/Southern Hemisphere	*Nothofagus*	28	1·9
15 Distribution types		1465	100·0

surveys in plant geography, the genus is the most practical taxon to use. Table 4.1 gives the names of the distribution types recognised by van Balgooy and also indicates a characteristic genus and the number of genera involved for each distribution type. It also shows the percentage contribution of each type to the total flora of New Guinea. A careful study of these figures clearly suggests that the flora of New Guinea is much more closely allied to that of Indo-**Malesia** than to that of Australasia.

The use of geographical elements is not confined, however, to the floristic analysis of geographical areas. Plant sociologists have long recognised their

Table 4.2 The spectrum of floristic elements (as defined by Fournier 1946) for 'L'association à *Scirpus fluitans* et *Apium inundatum* (*Scirpetum fluitantis*)' of the Marquenterre, Picardy, northern France (*after* Wattez 1968, 69).

Elements	% of species in the association
Cosmopolitan	30·6
Circumboreal	19·5
European	16·6
Atlantic	11·1
Old temperate and Eurasian	8·2
Eurosiberian	5·2

importance in characterising vegetation associations. For example Wattez (1968) examined the percentage presence of floristic elements – as defined by Fournier (1946) – in 'L'association à *Scirpus fluitans* et *Apium inundatum (Scirpetum fluitantis)*' of the Marquenterre in Picardy, Northern France, noting the significant presence of circumboreal and atlantic elements within this association (Table 4.2). His analysis was based on nine quadrats or *relevés* taken in the communal pastures of the commune of Larronville.

Yet, widespread as the use of geographical elements is, it is important to stress that there are many pitfalls to be avoided when assessing their value. The most dangerous is without doubt the often in-built assumption that the taxa comprising a particular element must all possess the same floristic history, emanating from the same centre of origin and governed by identical ecological constraints. This is rarely the case and great care needs to be taken when developing phytogeographical theories based on an analysis of floristic elements. Most elements are simply a recognition of similarity in the present-day distribution patterns as mapped and these patterns may tell us very little about the past distribution or the present status of the plant. Although today two plants may possess similar distribution patterns, one may be actually in the process of expanding its area, whereas the second may be suffering a contraction of area. Without a detailed study of all the taxa involved, it can be very dangerous to generalise.

Moreover, as Jardine (1972) is at pains to point out, there is a tendency for the human eye to discover groupings even where none is present and to select particular distributions as 'types', clustering the remaining distributions around them. For these reasons some recent work has made use of the processes of automatic classification (e.g. Proctor 1967), by which means it is hoped to achieve greater objectivity. The method has been succinctly outlined by Jardine.

An overall area is selected and partitioned into unit areas. Care is needed at this stage, for, if too coarse a grid is used, valid floristic elements may be agglomerated and lost. Arbitrary areas, such as the British Watsonian vice-counties (see pp. 36–8), also may be employed for this purpose. A measure of association between each pair of species distributions is then calculated. Jardine's suggestion is the number of squares in common/the number of squares in which either or both occur. Alternative measures of similarity are reviewed by Cormack (1971), Dagnélie (1960) and Goodman and Kruskal (1954). A cluster method is now applied to the matrix of similarity values. The choice of an appropriate cluster method is, however, a controversial matter (Jardine & Sibson 1971). Non-hierarchic cluster methods are, in principle, the most appropriate in the study of floristic elements but they tend to be computationally laborious. Moreover, they can only handle around 100 species. The usual hierarchic cluster methods, such as the single-link method, can deal with up to 2000 species. These methods produce what is termed a dendrogram, a hierarchy in which groups have associated numerical levels. The last stage in the process is the selection, by the investigator, of clusters which may be recognised as floristic elements.

Because cluster methods can sometimes find clusters even where the data

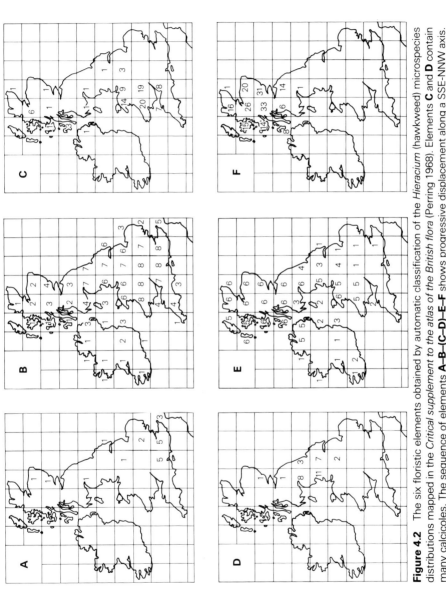

Figure 4.2 The six floristic elements obtained by automatic classification of the *Hieracium* (hawkweed) microspecies distributions mapped in the *Critical supplement to the atlas of the British flora* (Perring 1968). Elements **C** and **D** contain many calcicoles. The sequence of elements **A–B–(C–D)–E–F** shows progressive displacement along a SSE-NNW axis. Note the use of 100 km grid squares as the basic unit area. The numbers indicate the number of microspecies of that particular element in each grid square (*after* Jardine 1972, Fig. 1; © Linnean Society of London).

show no significant cluster structure, it is usually prudent to calculate a measure of the accuracy with which the system of clusters obtained represents the similarity matrix (Hartigan 1967, Jardine & Sibson 1971). If the system of clusters turns out to represent the similarity matrix very inaccurately, then a statistical test for cluster structure should be applied (Jardine 1971).

Figure 4.2 shows the distributions within Britain of six floristic elements obtained by automatic classification of the *Hieracium* (hawkweed) microspecies mapped in the *Critical supplement to the atlas of the British flora* (Perring 1968).

Isochores and isoflors. The identification of floristic elements is not, however, the only traditional method of studying the affinities between distributions of taxa. In his *Outline of the history of arctic and boreal biota during the Quaternary period* published in 1937, Hultén developed a theory that during the Pleistocene certain areas in higher latitudes had acted as refuges for some of the species of the pre-glacial flora, which had persisted at these sites until the return of more suitable climatic conditions and which had been able to respread subsequently from their refuges. To illustrate this thesis, he constructed a series of maps using the principle of isochores. These are cartographic lines, comparable with many other more familiar lines such as isotherms and isohyets, but in this instance linking places with similar species populations. Each successive line represents a higher proportion present of the total species of the particular group of species under study. In the case of Hultén's own work, these were the species which were thought to have respread from their isolated refuges. The idea is thus to detect in present distributions of species a trend to progressively wider areas of dispersal.

Unfortunately, it is first necessary in this approach to define the group of species in which one is interested and Raup (1941) regarded this as the main weakness of the method. Moreover, it should be noted that the lines do not depict the actual distribution of any one or more species. In fact Hultén's isochore maps were meant to be the cartographic expression of his 'Theory of equiformal progressive areas', in which he argued that plants of the boreal and arctic regions 'can be grouped around centra from which they must have spread, and that the total areas of all plants spreading from the same centre must form more or less concentric and equiformal progressive figures'. It is interesting to note that Daubenmire (1975) has recently proposed an opposite theory, namely one of 'equiformal recessive' ranges, for the Pliocene temperate mesophytic forest elements of the northern Rockies. Today these tend to share a coastal area, having vacated the Rockies to varying degrees. At their staggered eastward extremities the populations are highly disjunct and exhibit Daubenmire's equi-formal recessive ranges, for which he has defined a characteristic series. He says that 'hundreds of species found in eastern Washington and northern Idaho' can be matched with this series, thus 'confirming the reality of the mass exodus which resulted from changes in inland climates, which in turn followed the building up of the Cascade Mountains.' This change in climate involved in-creasing aridity, and the temperate zone mesophytes were eliminated from the lowlands of eastern Washington and northern Idaho, giving way to an influx of xerophytic species.

Isoflors, on the other hand, are used to locate centres of origin and migratory patterns in higher taxa, especially the genus. From the generic centre outwards, the number of species belonging to that genus may be expected to decrease regularly and to assume a pattern which suggests the tracts of past migration. Isoflors are cartographic lines linking regions with equal numbers of species belonging to the genus, or other taxonomic rank, in question. In the ideal expansion of a genus, the isoflors would be roughly concentric, but this rarely occurs in nature, the pattern being inevitably disrupted by ecological barriers and historical changes. By drawing these lines for the species within a genus, it may be possible to trace the generic centre, but a confident result can only be achieved when fully supported by palaeobotanical evidence (see Chs 5 and 6) and many groups have multiple centres of variation which greatly confuse the issue. Isoflor maps were used by Willis (1922) to illustrate his 'Theory of age and area' (see pp. 68–70).

The term isoflor was also employed by Chaney (1940a) for cartographic lines linking sites with similar types of fossil Eocene floras. The floras involved were designated in climatic terms as cool-temperate, temperate, intermediate and sub-tropical. Such isoflors can be valid only where there is no doubt concerning the contemporaneity of the fossil floras involved.

Floristic affinities between geographical areas

The traditional method of examining the floristic affinities between geo-graphical areas is the recognition of what are termed floristic units or floristic regions. The average geographical range of species is fairly small and at the world scale these limited ranges may be grouped geographically into regional units or floras which are differentiated on the basis of their distinctive taxa. The process is not an easy one because there is no universally agreed set of criteria for estimating the floristic differences between regions. Each plant geographer will devise his own guidelines to suit the particular purpose in hand. In practice these will probably involve some assessment of the nature and degree of endemism in floras, a consideration of the size and floristic importance of different floras and a survey of what van Steenis has called demarcation knots. These are points or lines of sudden change in floristic composition brought about by the presence of important geographical barriers or by the advancing front of a particularly aggressive flora. In his detailed work on the Malesian flora, van Steenis (1950) recognised three principal demarcation knots which are shown in Figure 4.3. Unfortunately, most boundaries are not easy to define and there is a wide area for personal interpretation and judgement.

Most modern attempts at a worldwide floristic arrangement, such as the well-known classification of Good (1974, 30–32 & Plate 4), are based on the original scheme of Engler (Engler & Diels 1936, 374–86) and vary from it in detail rather

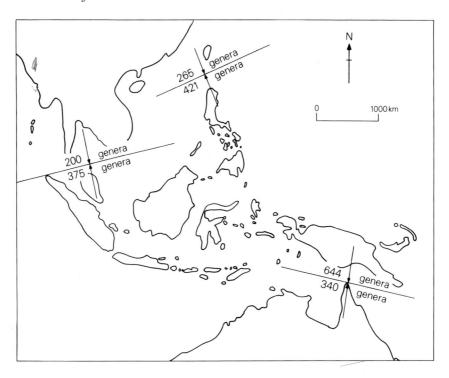

Figure 4.3   The three principal 'demarcation knots' of the Malesian flora (*after* van Steenis 1950; taken by permission from Whitmore 1975, Fig. 1.2). Note that Malesia is a modern term used by biogeographers to denote the islands on and between the Sunda and Sahul shelves. It was coined to avoid confusion with the State of Malaysia. Some biogeographers, however, prefer to draw the southern boundary at the base of the Papuan mountains and not, as here, in the Torres Strait (see Walker 1972).

than in principle. Frequently the regions are further divided into sub-regions, sectors or districts, and the regions themselves are usually grouped together into floristic kingdoms or empires, the largest units employed. Figure 4.4 is a map of the empires recognised by Lemée (1967). These were drawn up on the basis of both floral and faunal evidence.

Jardine (1972) points out that, just as cluster analysis can be applied to measures of similarity between species distributions to generate floristic elements, the same techniques can be used for measures of similarity in species composition to provide floristic regions. So far, however, most attempts to employ automatic classification of unit areas have been concerned with the study of faunal regions (e.g. Hagmeier & Stults 1964), although there is probably a wide scope in plant geography for the use of computer methods in the analysis of floristic affinities between geographical areas.

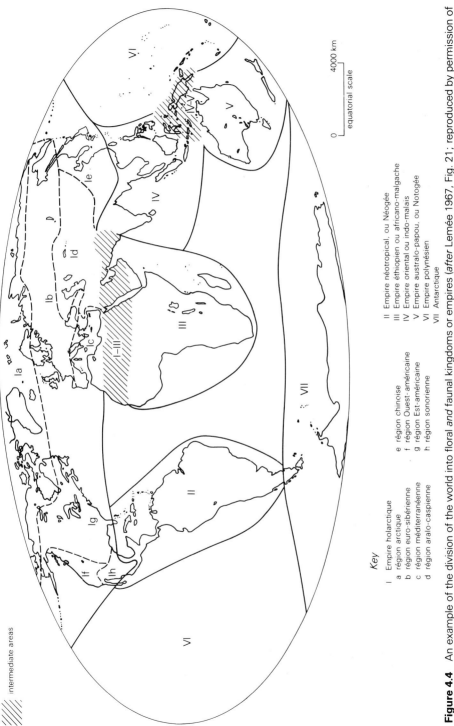

intermediate areas

Key

I Empire holarctique
  a région arctique
  b région euro-sibérienne
  c région méditerranéenne
  d région aralo-caspienne

  e région chinoise
  f région Ouest-américaine
  g région Est-américaine
  h région sonorienne

II Empire néotropical, ou Néogée
III Empire éthiopien ou africano-malgache
IV Empire oriental ou indo-malais
V Empire australo-papou, ou Notogée
VI Empire polynésien
VII Antarctique

Figure 4.4 An example of the division of the world into floral *and* faunal kingdoms or empires (*after* Lemée 1967, Fig. 21: reproduced by permission of Masson, Editeur s.a.). The main divisions recognised by Lemée may be given in English as: I Holarctic; II Neotropical or Neogaea; III Ethiopian; IV Oriental; V, Notogaea; VI, Polynesian; VII, Antarctic (cf. Illies 1974).

Types of natural distribution

So far in this chapter we have been concerned with methods of identifying natural groupings of taxa based on their geographical areas as mapped. It is now necessary to consider the shape or form of plant distributions to see if there are any frequently recurring areal patterns that can be widely recognised in more general terms as significant types of natural distribution.

At the very simplest of levels, all plant distributions fall into one of three fundamental types. These three types may be designated as wide or continuous distributions (wides), broken or disjunct distributions (disjuncts) and local or endemic distributions (endemics), the basic form of each type being depicted schematically in Figure 4.5. But these represent only the bare bones of a much more elaborate classification that will be developed further in later chapters of the book (see especially Chs 7 and 8). Two particularly important variations on the basic scheme are the recognition of vicarious areas and relict areas. Vicarious areas are those belonging to two very closely related taxa, which are possibly

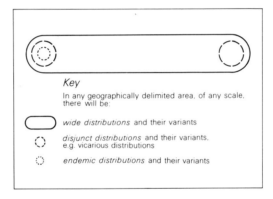

Key

In any geographically delimited area, of any scale, there will be:

wide *distributions* and their variants

disjunct distributions and their variants. e.g. vicarious distributions

endemic distributions and their variants

Figure 4.5   A schematic representation of the three basic types of plant distribution pattern. See the text for full explanation.

derived from the same common ancestor, but which now occur in separate geographical areas. This distribution type can be regarded as a variant of both the disjunct distribution (see pp. 104–5) and the endemic distribution (see pp. 115–16). Relict areas, on the other hand, are those exhibited by relict species (relicts) that now possess only a remnant of a formerly wide distribution. Again this is a variant of either the disjunct or the endemic type of distribution and it will be exemplified a number of times in later chapters. In the case of wides, it is frequently necessary to recognise a whole range of different categories at the world scale, such as the truly cosmopolitan wide, the tropical wide or the temperate wide, together with more specialised groups like the American wide or the Asiatic wide. All depends on the scale at which one is working.

The identification and interpretation of these various types of distribution are the subjects of the second part of this book.

Further reading

Daubenmire, R. 1975. Floristic plant geography of eastern Washington and northern Idaho. *J. Biogeog.* **2,** 1–18. (For Daubenmire's theory of 'equiformal recessive' ranges.)

Good, R. 1974. *The geography of the flowering plants,* 4th edn. London: Longman. (Ch. 2 & Plate 4.)

Jardine, N. 1972. Computational methods in the study of plant distributions. In *Taxonomy, phytogeography and evolution,* D. H. Valentine (ed.), 381–93. London & New York: Academic Press.

Jardine, N. and R. Sibson 1971. *Mathematical taxonomy.* New York: John Wiley.

Matthews, J. R. 1955. *Origin and distribution of the British Flora.* London: Hutchinson. (A classic study of geographical elements.)

Seddon, B. 1971. *Introduction to biogeography.* London: Duckworth. (Ch. 2.)

II INTERPRETING PATTERNS OF DISTRIBUTION

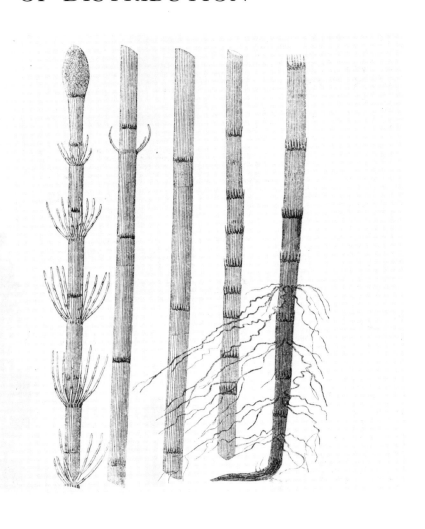

5 *Origins, boundaries and disruptions*

The interpretation of any plant area depends on a detailed understanding of the factors of distribution, both internal and external, which were described in Chapter 1. It is the purpose of this present chapter to analyse the significance of five of these factors, namely the centre of origin of a distribution (O), the ecological control of range boundaries, the age of taxa (t), and the phytogeographical importance of geological and climatic change. The other factors mentioned in Chapter 1, such as dispersal capacity (D), will be considered in relevant contexts in later chapters. The five factors highlighted here are either particularly problematic to study or are now recognised as of outstanding interest in the interpretation of plant ranges.

The centre of origin of a distribution

In theory, the geographical location of a taxon's centre of origin (O) should have a fundamental effect on the nature and orientation of a distribution pattern. Granted that many, though by no means all, taxa have only a limited dispersal capacity (D), the developing distribution is likely to cluster around the centre of origin for a long period of time after the moment of origin. The pattern is thus determined, in part, by the geographical coordinates of the centre of origin. If the plant had arisen elsewhere, its distribution pattern would obviously have been different. It might therefore be thought particularly important when interpreting a distribution to identify the centre of origin of that distribution.

The first systematic attempts to discover criteria for the determination of centres of origin were those of Adams (1902). These were developed and discussed at some length by Cain in a lecture given in 1942 to the Torrey Botanical Club, which was then published as Chapter 14 of his book *Foundations of plant geography* in 1944. In all, he considered thirteen criteria that might be used to indicate in some way the centre of origin of a distribution. As Adams wrote, such criteria should be regarded solely as 'convenient classes of evidence to which we may turn for suggestions and proof as to the origin and dispersal of organisms . . . . It should be clearly emphasised that it is the convergence of evidence from many criteria which must be the final test in the determination of origins . . . .'

According to the criteria considered by Adams and Cain, the centre of origin of a plant distribution *may* be indicated by:

(a) the location of the greatest variety of forms of the taxon;
(b) the location of the area of greatest dominance and density of distribution;
(c) the location of the most primitive forms;
(d) the location of the area exhibiting the maximum physical development of individuals;
(e) the location of the area of maximum ecological productivity of the taxon;
(f) continuity and convergence in the lines of dispersal;
(g) the location of least dependence on a restricted habitat;
(h) the identification of continuity and directness of individual variations or modifications radiating from the centre of origin along the highways of dispersal;
(i) the area of origin indicated by natural geographical affinities;
(j) the direction of origin indicated by the annual migration routes of animals, especially birds;
(k) the region of origin indicated by seasonal appearance or general **phenology**;
(l) an increase in the number of dominant genes towards the centre of origin (this criterion was first proposed by Vavilov in 1927);
(m) the concentricity of progressive equiformal areas (for a discussion of this criterion developed by Hultén, see Chapter 4, pp. 52–3).

As pointed out by Cain, many of these criteria are far from satisfactory, especially criteria (d), (e) and (g), which are much more concerned with ecologically induced variation than with strictly distributional questions about the evolutionary centre of origin. Indeed, all the criteria proposed are subject to serious problems of interpretation and must be used in conjunction for any sense of reliability to be achieved.

For an example of a recent attempt to apply some of these criteria in the identification of a centre of origin, we may refer to the work of Holland (1978) on the evolutionary biogeography of the genus *Aloë*. On the assumption that all 350 species of *Aloë* have evolved from a common ancestor, Holland has applied Cain's first criterion to try to locate the centre of origin of the genus. This criterion states that the centre of origin should possess a greater variety of forms of the taxon than one might expect to find in areas near to the edge of the range. Two criticisms can be made of this argument. First, habitat diversity may lead to the development of large numbers of closely related forms well away from the true centre of origin. Secondly, considerable morphological differentiation can result from the later formation of **polyploid** complexes (see Ch. 9).

Holland restates Cain's criterion in statistical terms. Assuming that it is possible to arrange the different forms of a taxon in an evolutionary sequence, one can then assign an arbitrary number value to each. 'The centre of origin of a taxon, possessing the greatest variety of forms and a concentration of the more primitive of them, should', he argues, 'be indicated by the largest value for standard deviation and smallest for the arithmetic mean. Conversely, places near the edge of the range should be typified by larger means ($\bar{x}$) and smaller standard deviations (s).' He then proposes an index of dispersion ($s/\bar{x}$) and

Table 5.1 Assumed evolutionary classes in the genus *Aloë* (*after* Holland 1978, 223).

			% of all species
Least advanced	Class 1	small herbaceous plants with narrow linear leaves	7
	Class 2	shrubby plants normally without stems	47
	Class 3	plants shorter than 1·5 m at maturity and with unbranched stems	18
	Class 4	branched, somewhat shrubby plants typically not exceeding 1·5 m at maturity	15
	Class 5	arborescent plants usually much taller than 1·5 m and with unbranched stems	8
Most advanced	Class 6	tall branched arborescent plants	5

applies this to the genus *Aloë*. It will be noted that Holland's approach actually links Cain's criterion (a) with his criterion (c), the location of the most primitive forms.

The evolutionary sequence drawn up by Holland for the genus *Aloë* accepts that the **herbaceous** forms are less advanced than the arborescent (Table 5.1). The various species of *Aloë* found native in Africa are then numerically scored according to his system. The arithmetic means, standard deviations and indices of dispersion are given in Table 5.2. The origins of the genus should be indicated by the largest value for the index of dispersion.

The values shown would suggest that the ancestral aloes first appeared in south-east Africa sometime before land connections with Malagasy were severed in the late Mesozoic – early Tertiary (Fig. 5.1). From there, it seems that they may have dispersed along the rising highlands of eastern and southern Africa, reaching the Arabian Peninsula by about the latter part of the Tertiary. Within the present distribution, it is possible to detect some eleven or twelve

Table 5.2 Statistics for the morphological–evolutionary status of aloes in areas of unusually high species density (*after* Holland 1978, Table 1; reprinted by permission of Blackwell Scientific Publications Ltd).

	No. of species	Arithmetic mean $(\bar{x})$	Standard deviation (s)	Dispersion $(s/\bar{x})$ %
Arabian Peninsula	20	2·9	0·9	31
Somalia	18	3·2	1·4	43
Ethiopia	17	2·9	0·8	28
East Africa	35	3·0	1·1	37
Malawi–Zambia	14	2·5	1·4	56
Rhodesia	24	2·3	1·4	61
Transvaal	66	2·7	1·3	48
Malagasy	45	2·7	1·2	44
Eastern Cape	29	3·3	1·4	42
Namibia	20	3·5	1·3	37
Angola	20	3·0	1·3	43

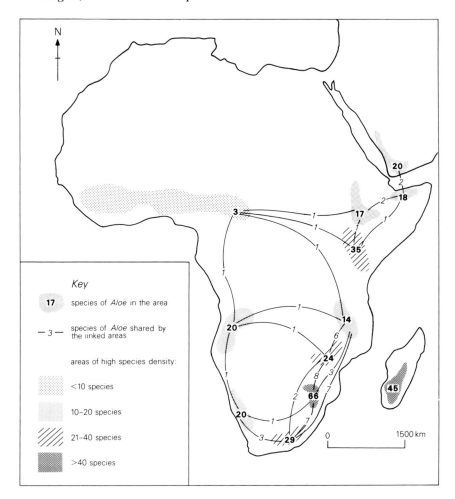

Figure 5.1 The genus *Aloë:* areas of unusually high species density and their taxonomic affinities (*after* Holland 1978, Fig. 3; reproduced by permission of Blackwell Scientific Publications Ltd).

areas which represent secondary centres of variation or frequency. In these regions, new taxa probably arose through the process of adaptive radiation, which is the diversification of form in response to the pressures of different ecological habitats. It is clearly important in discussing any distribution pattern to distinguish carefully such secondary centres of variation, frequency and development from the true centre of evolutionary origin.

So far in our discussion, it has actually been assumed that it is feasible to locate a single geographical centre of origin. For some taxa, this may not be so. No single point of origin would, for example, exist for a species arising by **chromosome** doubling which occurred many times and in many places throughout the progenitor **diploid** population. Moreoover, for many taxa the true centre of origin has been lost, that part of their distribution having been wiped out by the external forces of geological change, climatic change or the actions of man. Thus, although in theory the identification of a taxon's centre of origin is desirable, in practice it is a difficult and often impossible task. In this particular aspect of plant geography, therefore, there will always be a great need for 'inductive reasoning, with assumptions reduced to a minimum and hypotheses based upon demonstrable facts . . .' (Cain 1944, pp. 210–11).

Boundaries

Of the internal factors governing plant distribution, the most important factor controlling the final character of the boundary of that distribution is, without doubt, the autecology (P) of the plant in question. Granted that the plant has the time (t) and the dispersal ability (D) to occupy all available habitats, it will be the distribution and shape of these habitats which will determine the pattern of the plant's geographical range. Obviously a change in the geology, climate or some other attribute of the habitats will, in turn, alter the ecological framework within which the plant's autecology is expressed and thus the boundaries and the topography of the distribution. Plant geographers are therefore especially concerned to understand as fully as possible the exact nature of these ecological controls and it is often at the boundaries of a distribution that they are most evident, for here the taxon tends to be under ecological stress. The kinds of

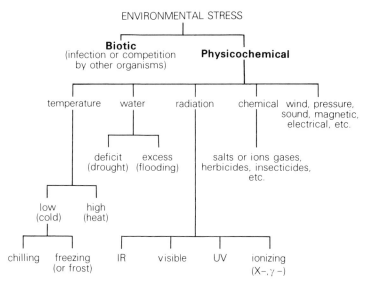

Figure 5.2 The kinds of environmental stresses to which an organism may be subjected (*after* Levitt 1972, Fig. 2.2; reprinted by kind permission of the author and of Academic Press Inc.).

stresses involved are shown in Figure 5.2 and the general subject of environmental stresses has been reviewed in detail by Levitt (1972).

Unfortunately we know surprisingly little about the autecology of most
species. The factors controlling the edges of a range appear to be very complex
and can be difficult to analyse. Climate and soil are obviously the main
determinants, but the rôle of **competition**, fire, disease or animal predation
may be just as important. In very few cases will the position of a boundary be
determined by one factor alone, although, for example, a sudden and rare frost
may significantly alter the boundary of a frost sensitive species.

Towards the edge of its natural climatic range, a plant often displays a
reduction in its reproductive and dispersal capacity. Sterility is frequently a first
sign that a species is near to its effective limit of distribution. To reproduce
successfully, a plant may well require higher temperatures or a longer period of
optimal temperature than obtain at the margins of its area, where, however, it
can still maintain its vegetative functions. This is well demonstrated by the work
of Iversen (1944) who developed the idea of the thermal curve, a topic discussed
at length by Seddon (1971). To obtain this curve for any taxon, the mean
temperature of the warmest month is plotted as the ordinate and that of the
coldest month as the abscissa for climatic stations near to the boundary of the
taxon in which one is interested. This produces a scatter diagram. The presence
or absence and the reproductive performance of the plant within the vicinity of

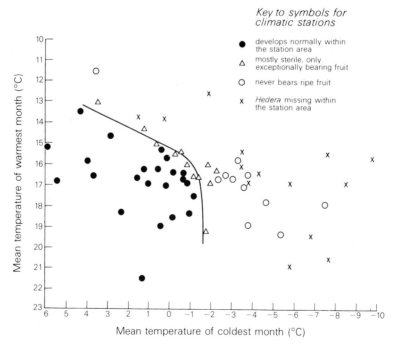

Figure 5.3 Thermal limits and the thermal curve at the boundary of distribution of the ivy
(*Hedera helix*) (*after* Iversen 1944 and taken with permission from Seddon 1971, Fig. 4.2).

these stations is then assessed and suitable symbols superimposed on the scatter diagram to indicate the differences for each locality (Fig. 5.3). On the graph, these symbols will clearly distinguish sites at which the plant is present from those where it is not and also differentiate the stations where its reproductive capacity is reduced.

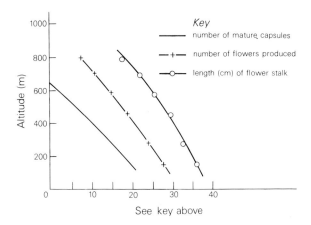

Figure 5.4 The effect of altitude on certain characteristics of the heath rush (*Juncus squarrosus*) (*after* Pearsall 1950).

As with the example of ivy (*Hedera helix*) given in Figure 5.3, the stations showing reduced fertility will probably lie approximately between the regions of presence and the regions of absence indicating a true boundary zone. A line drawn through these points will thus define the thermal limits of the plant's distribution in terms of both summer and winter temperatures. The thermal curve of *H. helix* indicates that it is very intolerant of cold winters and is unable to produce viable seed where the coldest month has a mean temperature below −1°C. Summer warmth, however, is also important, and where no summer month has a mean temperature greater than 16°C, its tolerance of winter cold is even less. Thus, as Seddon notes, in the mildest of temperate winter conditions (monthly mean 4·5°C), *H. helix* is limited by inadequate summer warmth. From this case alone, it is plainly clear that the response to climate is very complex and that quite different factors will be effective in controlling the boundaries of taxa.

Altitudinal boundaries are also of great interest to the plant geographer and likewise tend to reflect subtle climatic changes. Pearsall (1950), for example, has shown that the heath rush (*Juncus squarrosus*) exhibits a marked decline with altitude in both the number of flowers and the number of mature capsules produced (Fig. 5.4). This trend appears to reflect the changing length of the warm season. In 1942, flowering was completed at 215 m elevation during June, but at 610 m flowering did not begin until the end of July and at 760–915 m it was not completed until the end of August.

The study of plants at their boundaries is thus a fascinating challenge. Too little has been done on the detailed autecology of species and many generalised explanations of range limits leave much to be desired. If two plants have similar distribution boundaries, it does not automatically follow that these are determined by the same ecological factors.

Time

To many, the proposition that the area (A) occupied by a species is in some way proportionate to its age, which is the time that the species has existed, may seem axiomatic and yet simplistic. Granted an unchanging world, a plant distribution would evolve at a rate commensurate with its capacity to disperse and establish itself and the process of development would cease only when all suitable habitats had been occupied. Until that time, the size of the distribution area would undoubtedly reflect how long the taxon had been in existence. The older it was, the longer it would have had to disperse itself. Obviously, if two species arose at the same time, but one possessed a far more efficient dispersal system (D), then it is likely that the more efficient taxon would be occupying a larger area at any given time, assuming that its spread was not restricted by ecological constraints (P).

But the real world is not so simple. As we have already seen in the discussion on centres of origin, few distributions are left unchanged by external factors. The basic progression from the centre of origin to the complete distribution is disrupted by geological, ecological and biological events and a formerly wide-spread taxon may have its distribution reduced to a few enclaves, so that its area of distribution is far smaller than that of many younger taxa. The relationship between age and area is therefore a great deal more complex.

An understanding of this relationship between time and area was implicit in the writings of many early plant geographers and geologists, such as Lyell (1853) and Hooker (1853), but its fullest expression came in the work of Willis (1922, 1940, 1949) and with his 'Theory of age and area'. Willis regarded his theory as a rule, a general law 'covering all or nearly all the plants now existing upon the globe'. In 1922, the rule was expressed thus:

> The area occupied at any given time, in any given country, by any group of allied species at least ten in number, depends chiefly, so long as conditions remain reasonably constant, upon the ages of the species of that group in that country, but may be enormously modified by the presence of barriers such as seas, rivers, mountains, changes of climate from one region to the next, or other ecological boundaries, and the like, also by the action of man, and by other causes.

The geographical expression of this theory is suggested by Willis' map showing the areas occupied by the species of the genus *Haastia* in New Zealand

(Fig. 5.5). According to the 'rule of age and area', the taxa with the smallest areas would be the youngest in the genus.

Willis and his theory have received a great deal of often bitter criticism (e.g. Berry 1924, Gleason 1924, Good 1974, Greenman 1925, Schonlands 1924, Sinnott 1924). Some of this he heaped upon his own head by making his theory too wide and all-embracing, and by his adherence to a mathematical treatment of the concept which involved a kind of graph known as the hollow curve.

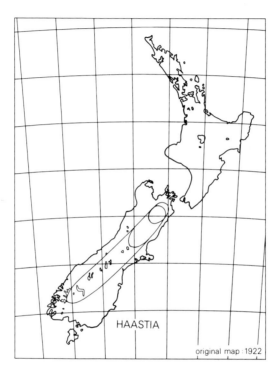

Figure 5.5 A map showing the areas occupied by certain groups of species of *Haastia* in New Zealand. This was one of the original maps used by Willis (1922, 154) to illustrate his Theory of age and area (reproduced from J. C. Willis, 1922, *Age and area* by permission of Cambridge University Press).

Unfortunately, this did not really work, and it certainly alienated many botanists from his ideas (Nicholson 1951). Moreover, the theory is subject to many caveats, including the fact that it is perhaps too closely related to Willis' distinctive views on evolution (for these, see Willis 1940) and the fact that it clearly demands circumstances in which evolution has proceeded under uniform conditions for very long periods.

Nevertheless, as Good (1974) points out in his clear survey of the theory, 'Willis did the cause of plant geography a great service by publishing it'. In the

1920s, it helped to revitalise interest in the theory of the subject and it certainly corrected an imbalance that had grown up in plant geographical thought about the importance of the time factor. As Willis himself wrote, 'the simple effects of age upon dispersal' had 'been lost sight of, under the widely held view that distribution was rapid, and that local species were either local adaptations or were dying out.' His books are still well worth reading.

Disruptions

In both Chapter 1 and the preceding discussion on centres of origin, boundaries and the significance of age in relation to area, it has been repeatedly emphasised that most plant distributions are the product of not only the internal controls (I) of distribution but also of a wide range of outside influences for change (E). There can be no perfect model for plant distribution in which an uninterrupted process leads to an inevitable conclusion. Indeed, external changes may so completely obliterate the original distribution that any new pattern bears no relation whatsoever to the previous one.

Recent research has greatly underlined the phytogeographical importance of two such external forces for change. The first of these is geophysical in character and involves the old concept of continental drift and the newer theories of sea-floor spreading and global plate tectonics. The second concerns the reality of climatic change and the influences of such changes on the distribution of plants. Both these topics are now considered in some detail.

The Earth's moving surface

The idea of moving continents or continental drift is in fact an old one. As early as 1658, a Frenchman called François Placet had suggested that the Old and New Worlds might have become separated following the Biblical Flood. In 1858, Antonio Snider-Pelligrini produced the first diagram to fit together the continents bordering the Atlantic and in 1861, Pepper invoked the same concept to explain the occurrence of identical fossil plants in the coal deposits of both North America and Europe. Early in the twentieth century two Americans, Taylor and Baker, gave further, and elegant, expressions of the theory, but it was with the publication of a book called *Die Entstehung der Kontinente und Ozeane* by the German geophysicist, astronomer and meteorologist, Alfred Wegener, that the modern theory of continental drift was truly launched. Although this work first appeared in 1915, it was the 1924 English translation of the 1922 German edition that really started the debate. The story since then has been one of continuing discovery of more and more evidence in support of moving continents and by the late 1960s scientific opinion had without doubt swung towards the acceptance of the theory.

It is now, however, a very different theory. No longer are we simply concerned with moving continents, but with two new and far more comprehensive concepts, those of sea-floor spreading and plate tectonics. These important geophysical theories are fully explained in many excellent texts, to which the reader is referred (e.g. Cox 1973, Hallam 1973, Le Pichon *et al.* 1973, Phinney 1968, Runcorn 1962, Tarling & Runcorn 1972, Tuzo Wilson 1972, Windley 1977). A readable and easily understood introduction to the subject is provided by Tarling and Tarling (1972).

From the point of view of plant geography, the significance of the new theories is twofold. First, they are attested to by at least four main lines of independent geophysical evidence which have nothing whatsoever to do with the study of plant distributions. The dangers of a circular argument in which distinctive plant distributions are used to suggest continental movement, which is then in turn employed to interpret the distributions, are thus avoided. The theory of movement is provided by a different science entirely and can therefore be rightly used as a tool of interpretation. As Cox (1973) indicates, the new synthesis has arisen through evidence gleaned from:

(a) mapping the topography of the sea floor using echo–depth sounders;
(b) measuring the magnetic field above the sea floor using proton–precession magnetometers;
(c) timing the north–south changes of the Earth's magnetic field using both the magnetic memory and the radiometric age of rocks from the continents;
(d) determining very accurately the location of earthquakes using a worldwide net of seismometers originally developed to detect nuclear blasts.

Secondly, the theories present a whole new framework of thinking for plant geographers, which is only just being worked out. We are no longer interested solely in continents, but in a whole series of phenomena, such as mid–ocean ridges and island arcs, which pose totally different questions about, say, the relationships of drift and dispersal as alternative explanations of current disjunct distribution patterns (Ch. 7). Short-range dispersal, in stages, may have persisted for a long time between separating continents, for example. We must therefore re-think the application of the idea of moving continents in plant geography.

The basis of the new theories lies in the identification of six major and some fourteen minor lithospheric plates which are bordered by tectonic belts (rises, trenches and transform faults) (Fig. 5.6). Although some of these plates are solely oceanic in character, most consist of both continental and oceanic parts. The reason for this combination within a single plate is that new oceanic crust has been attached to old continental crust by the process of sea-floor spreading.

By the late Carboniferous, it is believed that all the continents had drifted together to form a continuous land mass that is usually termed Pangaea (meaning 'all lands'), a concept originally proposed by Wegener (Fig. 5.7).

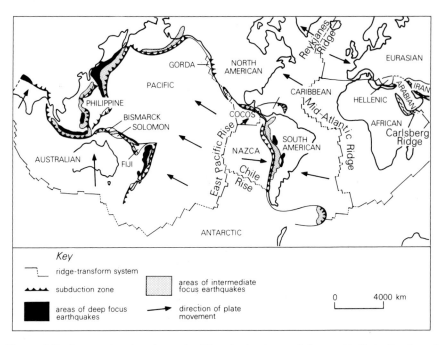

Figure 5.6 A map showing the major lithospheric plates of the world, their direction of movement and the relationship between their boundaries and the seismic belts of the world (*after* B. F. Windley 1977. *The evolving continents*. Fig. 15.1. © 1977 by John Wiley & Sons Ltd. Reprinted by permission of John Wiley & Sons Ltd).

Other theorists are unhappy with the idea of such an all-embracing land mass and prefer to reconstruct the two later supercontinents of Laurasia and Gondwanaland, a division suggested by du Toit as far back as 1937. Such a division appears to have obtained in any case by the early Jurassic, some 180 million years ago (Fig. 5.8). The breakup of Pangaea continued throughout the Mesozoic and the Cenozoic, with the last continent separating in the Tertiary about 45 million years ago (Figs 5.9 & 5.10). The reason for the breakup of Pangaea and the separation of the continents lies in the fact that the intervening ocean floor began and continued to grow, causing the drift of the continental plates. As lavas form new ocean floor, spreading laterally from ridges, the lighter, buoyant continents are rafted to new positions at rates of several centimetres per year.

This concept of ocean-floor spreading goes as far back as the work of Holmes (1931) but received its first fully cogent expression in the now classic paper of Hess (1962) entitled 'History of ocean basins'. Working in marine geology, Hess (Fig. 5.11) was undoubtedly the major figure in the development of the theories of ocean-floor spreading and plate tectonics.

The importance of these ideas for the interpretation of disjunct or broken intercontinental distributions is discussed later in Chapter 7 (see pp. 99–103). For the moment it will suffice to illustrate their possible significance in phytogeography

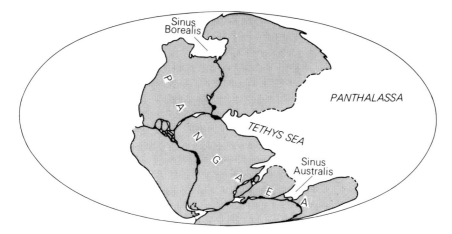

Figure 5.7 200 million years ago. A universal land mass, Pangaea, may have looked like this, with Panthalassa as the ancestral Pacific Ocean and the Tethys Sea (the ancestral Mediterranean) forming a large bay separating Africa and Eurasia (*after* Dietz & Holden 1972, 106).

by describing Raven and Axelrod's (1972) thoughts on the likely rôle of plate tectonics in the story of Australasian palaeobiogeography. These writers of course assume that the evidence from plate tectonics is basically correct.

During the middle Cretaceous, when Australia–Antarctica was still in proximity to Africa, India and New Zealand, before the final disruption of Gondwanaland, Raven and Axelrod envisage ties between the Australian, Antarctic and American plates which provided migration routes for species of the southern temperate forests (such as members of the Chile pine family

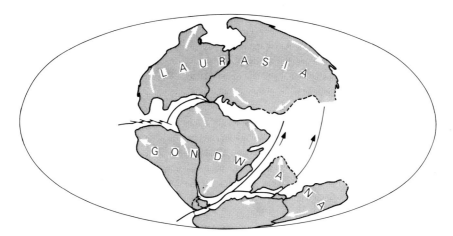

Figure 5.8 After 20 million years of drift. At the end of the Triassic period, 180 million years ago, the northern group of continents, known as Laurasia, has split away from the southern group, called Gondwana (*after* Dietz & Holden 1972, 107).

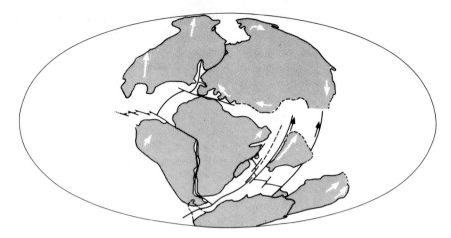

Figure 5.9   After 65 million years of drift. At the end of the Jurassic period, 135 million years ago, the North Atlantic and the Indian Ocean have opened up considerably and the birth of the South Atlantic is imminent (*after* Dietz & Holden 1972, 108).

(Araucariaceae), the Proteaceae, the Winteraceae and the southern beech genus *Nothofagus*). These trees persisted until later Eocene times. Then, as the Australian plate rafted north, it encountered the west-moving Pacific plate and underwent fragmentation, which resulted in the development of a complex series of basins, plateaux, island arcs and deeps. The isolated lands now provided refugia for the survival of ancient taxa, some of which also persist in distant regions such as Madagascar and Chile. In addition, new taxa evolved over the lowlands as they were rafted north to warmer climates. In Australia for

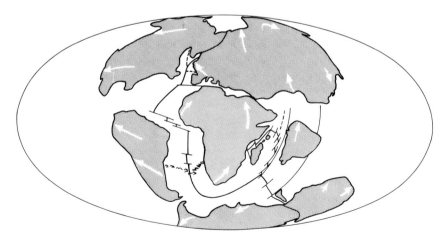

Figure 5.10   After 135 million years of drift. At the end of the Cretaceous period, 65 million years ago, the South Atlantic has widened into a major ocean and Madagascar has been separated from Africa (*after* Dietz & Holden 1972, 109).

example, as the plate moved into lower latitudes, the mixed Cretaceous–Eocene forests were replaced progressively by taxa adapted to mediterranean and desert climates, that is **xeric** shrublands and low woodlands, and savanna and tropical forest.

When the Australian plate collided with the Asian plate in the Miocene, the mixing of Oriental and Australian taxa began. In the late Cenozoic, the elevation of high mountains throughout the region from Malaysia to New Guinea and Australia–New Zealand provided new dispersal routes for temperate plants from the north (e.g. *Veronica* spp., *Euphrasia* spp., *Poa* spp. and *Carex* spp.) and from the south (*Astelia*), and new sites for their rapid evolution.

Figure 5.11   Harry Hess, the marine geologist who laid the foundations of the modern theory of sea-floor spreading. (Taken by kind permission from A. Cox, *Plate tectonics and geomagnetic reversals*. © 1973 W. H. Freeman & Co.)

In this one case alone, therefore, plate tectonics have provided a totally new framework within which the biogeography of a region can be interpreted. Although some workers remain critical of such a thesis, it is now clear that geophysical theories of plate tectonics and sea-floor spreading can never be ignored by plant geographers.

Climatic change

A second external factor also known to have disrupted the distributional patterns of organisms throughout geological time is climatic change. As we have already seen, the boundaries of most plant distributions are controlled by a complex series of climatic factors and it is obvious that any change in these factors will cause a taxon to expand or contract its area, depending on the nature of the change. Warmth-loving species will be excluded from regions during periods of colder climates but may well return when the climate ameliorates. The taxa present in any given area will therefore change as the climate changes. In some cases, however, species from one climatic regime may survive as relicts during a later period with a markedly different climate. Such relicts tend to persist on sites which provide localised refuges for the ecological conditions that suit their autecology (P). This subject of relicts and refugia is explored further in the next three chapters.

Perhaps the best-known evidence for the reality of climatic change is that for the successive advances and retreats of immense ice sheets over temperate latitudes during the past million years or so. The record of these Quaternary climatic changes forms the subject of many books (e.g. Butzer 1964, Embleton & King 1968, Evans 1971, Flint 1971, Washburn 1973, West 1977 and Wright & Moseley 1975) and is not repeated here. It should also be noted, however, that there is considerable evidence for a whole series of ice ages throughout geological time (Fig. 5.12) and these are discussed by Tarling (1978). It is not surprising, therefore, that the application of palaeobotanical techniques, which is discussed at length in the next chapter, has revealed significant vegetation changes taking place in response to these climatic changes. For the Quaternary such changes are documented for the tropics (Flenley 1979), Africa (e.g. Livingstone 1975), South America (Van der Hammen 1974), North America (e.g. Wright 1971) and Asia (e.g. Frenzel 1968, Tsukada 1966), as well as Europe. The subject of climatic change thus leads on naturally to a consideration of the evidence available for the reconstruction of past distribution patterns.

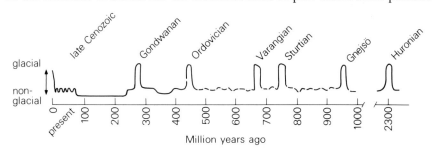

Figure 5.12   Ice ages through geological time. The peaks of the curve correspond to periods of major ice sheet formation with each period including several glacial and interglacial stages. The troughs correspond to periods with no known glaciation, with intermediate positions indicating the possible extent of mountain glaciation (*after* D. H. Tarling 1978. The geological–geophysical framework of ice ages, Fig. 1.1; reproduced from J. Gribbin (ed.), 1978, *Climatic change,* by permission of Cambridge University Press).

Further reading

Useful introductory texts are marked with an asterisk.

Gribbin, J. (ed.) 1978. *Climatic change*. Cambridge: Cambridge University Press.
*Heather, D. C. 1979. *Plate tectonics*. London: Edward Arnold.
Holland, P. G. 1978. An evolutionary biogeography of the genus *Aloë*. *J. Biogeog.* **5,** 213–26.
Keast, D. 1971. Continental drift and the biota of the southern continents. *Q. Rev. Biol.* **46,** 335–78.
Levitt, J. 1972. *Responses of plants to environmental stresses*. New York & London: Academic Press.
Raven, P. H. and D. I. Axelrod 1972. Plate tectonics and Australasian paleobiogeography. *Science* **176,** 1379–86.
Raven, P. H. and D. I. Axelrod 1974. Angiosperm biogeography and past continental movements. *Ann. Missouri Bot. Garden* **61,** 539–673.
*Seddon, B. 1971. *Introduction to biogeography*. London: Duckworth. (Chs 3 & 4.)
*Tarling, D. H. and M. P. Tarling 1972. *Continental drift*. Harmondsworth: Penguin. (First published by Bell, 1971.)
*Tuzo Wilson, J. (ed.) 1972. *Continents adrift* (Readings from *Scientific American*). San Francisco: W. H. Freeman.
Windley, B. F. 1977. *The evolving continents*. New York: John Wiley.

6 *Evidence for patterns of distribution from the past*

Although the plant geographer is primarily concerned with describing and explaining the present-day distribution of taxa over the world's surface, it is soon obvious that he is unable to do this satisfactorily without recourse to evidence from the past. The same distribution form or pattern is not always the product of the same processes. The currently restricted range of one taxon, for example, may represent either a distribution that was formerly extensive but which is now in retreat or precisely the opposite, namely the distribution of a young taxon which is in the process of expanding its geographical area. On the other hand, this same distribution may have been always a narrow one, naturally restricted by ecological barriers or the inherent inability of the taxon to spread further. Without detailed evidence concerning the past character of the distribution, attempts to interpret its present status may prove merely speculative.

Moreover, the plant geographer is also concerned with the changing nature of floras and floristic regions and the former distributions of key floristic elements. He will wish to relate these altering patterns to the geological and climatic changes discussed in Chapter 5, as well as to the activities of man and to the manner in which taxa evolve and spread. The study of past floras is therefore central to a plant geographer's work.

In all these matters, however, it is essential to avoid one very dangerous temptation, that of the circular argument. The present is too easily seen as the key to the past and then the past as the key to the present. For example, a particular floristic change may be invoked as evidence for a certain climatic alteration which itself is then used, in a later context, to account for the original floristic change. Great care is therefore needed in the way that evidence is marshalled and employed to generate and test theories.

Throughout this chapter we shall investigate the methods available for reconstructing past distribution patterns. These methods fall broadly into two categories. In the first, and most important category, the picture of the past is built up from the study of palaeobotanical and historical evidence. Although the techniques employed are subject to many problems of interpretation, they do tend to provide the most unambiguous evidence for the former presence or absence of a taxon in a given site or locality. In the second category, the main approach is what I will term the 'retrospective method'. Here the present is the key to the past and past distributions are reconstructed from the evidence of present-day patterns. The most thorough studies of course are those using both categories of evidence.

Palaeobotanical evidence

Palaeobotanical evidence for the past distribution of plants depends above all on the happy chance that many groups of plants are found fossil with their harder parts preserved. In practice two broadly different types of fossil are recognised and these require different analytical techniques in their study. The first group comprises the macroscopic fossils or macrofossils. These are large enough to collect and study with the naked eye or under low magnifications. The second are the microscopic fossils or microfossils. In the main these require special methods of extraction from the materials in which they are fossilised and are studied with the aid of a high-power microscope.

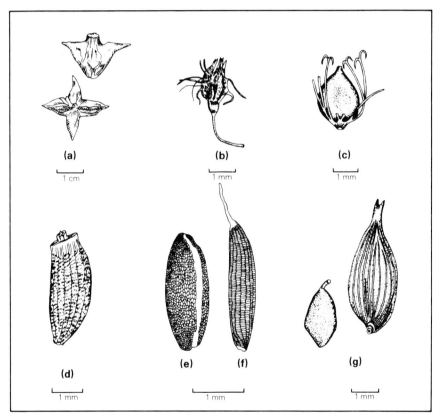

Figure 6.1   Some examples of macroscopic plant remains. (a) Nut of *Trapa natans,* a thermophilous water plant, found in interglacial and postglacial deposits. (b) Fruit of *Rumex maritimus,* a dock of damp places. (c) Fruit of *Polygonum lapathifolium,* a weed of waste places and damp ground. (d) Achene of *Cirsium arvense,* a thistle of pasture and waste places. (e) Fruit of *Najas flexilis,* and (f) of *Najas minor;* these are common plant remains in lake deposits. The former is found under a wide range of climatic conditions, temperate and cold; the latter is confined to temperate interglacial deposits. (g) The nut and fruit (nut within utricle) of *Carex riparia,* a fen sedge (*after* B. W. Sparks & R. G. West 1972. *The Ice Age in Britain,* Fig. 6.1; reprinted by permission of Methuen & Co Ltd).

Evidence from plant macrofossils. The most commonly recognisable macro-fossils are wood fragments, leaves, fruits and seeds (Fig. 6.1). Methods for their study, extraction and counting are given in the *Handbook of paleontological techniques* (Kummel & Raup 1965). Two main problems are involved in their use as evidence. The first is their identification and the second is the difficulty of discovering whether they were actually growing at or near to the site where they are now found fossil.

Many deposits, known as allochthonous deposits, were derived from distant and varied sources. In these cases, plant remains from different regions may have been collected in the one deposit, and it often proves impossible to separate the local from the alien. Accordingly, West (1977) has emphasised the need to study the taphonomy of fossil assemblages, which means the processes by which living communities are turned into fossil groupings. Figure 6.2 summarises these processes schematically and it is readily seen that the dispersal of plant remains depends on a series of biological and physical systems that lead to the final deposition of the fossil assemblage. Some of these systems may involve movement through, for example, soil creep, or river flow, or ocean currents, and plant remains may be transported great distances before they finally come to rest.

Moreover, subsequent weathering or erosion and the reworking of a formerly stable deposit can further disrupt the assemblage and redistribute its fossils, which then become derived fossils. This is in contradistinction to primary fossils which are those remaining in their original position of fossilisation.

Birks (1973) has pointed out that this need to understand more fully the processes of macrofossil formation and the dispersal of plant remains has resulted in a number of recent investigations into the ways in which present-day plants and vegetation are being turned into fossils. McQueen (1969), for example, has studied the dispersal of modern plant remains in a lake in New Zealand. He attempted to assess the significance of the transport of plant remains by streams flowing into the lake from higher altitudes. On the other hand, Ryvarden (1971) has surveyed the current dispersal of plant remains on to recently deglaciated ground in Norway. Birks herself has investigated the relationships between modern macrofossil formation and the living vegetation of small lakes in Minnesota. Her work emphasises the great caution that is necessary in interpreting macrofossil assemblages. It appears that the relationship between the distribution of the living plant and the distribution of its macro-fossil is nowhere near as simple as one would have hoped. Macrofossils can occur where the living plant is not growing, but need not occur where it is and may be more common in habitats significantly different from the habitat of the living plant. All this illustrates the importance of understanding more fully the processes of fossilisation.

If we consider the taxonomy of macrofossils, two further problems are encountered. The first involves the difficulties of identifying taxa from only partial remains, for it is very rare indeed to find a complete plant fossilised. The palaeobotanist is therefore working usually with two or three fruits, or a few leaves, or only part of a leaf. In many cases it is not feasible to identify the whole plant from such small detached parts, and in others confusion is inevitable. The second problem is that there is a difficulty in relating fossil taxa to the taxonomy of living species. This is essential if we are to interpret the phytogeographical and ecological significance of fossil floras, yet it is far from easy even if there are comprehensive macrofossil cross-reference collections. The problem is well illustrated by the frequent reporting of fossils from the northern hemisphere, which are claimed to belong to genera and species now fairly rigidly confined to

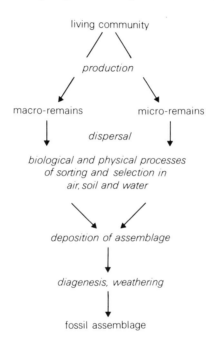

Figure 6.2 The processes involved in the production of fossil assemblages from living communities (*after* R. G. West 1977. *Pleistocene geology and biology,* 2nd edn, Fig. 7.7; reproduced by permission of Longman Group Ltd).

the southern hemisphere, such as the gum genus (*Eucalyptus*), the southern beech genus (*Nothofagus*), and members of the Proteaceae (for critical comments on such records, see Kausik 1943, Couper 1960, Good 1974). Careful investigation of these claims invariably shows that there is no completely non-controversial fossil among them. On purely morphological grounds, their identification is open to reasonable doubt and they are often accepted or rejected on somewhat arbitrary criteria. The wider application of the more advanced taxonomic research methods described in Chapter 2, such as electron microscopy, may help to resolve some of these enigmas.

Evidence from plant microfossils. In the main, plant microfossils comprise the pollen grains of higher plants and the spores of lower plants, although such matter as leaf hairs and epidermis and the skeletons of diatoms and desmids are also important. The two key properties of pollen grains and spores which make them so significant as microfossils are their great morphological diversity, which facilitates identification, and the resistant nature of their wall, which promotes their survival in sediments. This wall is formed of a complex carbohydrate (and possibly a protein element as well) that is chemically resistant, except to oxidation, and this means that pollen grains and spores can persist for hundreds of millions of years in unweathered sediments.

Such fossil pollen grains and spores are derived primarily from the air by a process of 'fall-out', known as pollen rain, whilst the skeletons of diatoms and desmids, which are common in marine sediments, originate in the water body depositing the sediment. As with macrofossils, it is vital to understand as fully as possible the processes which govern the production of the fossil assemblage if we are to be able to interpret its phytogeographical and ecological signficance. In the case of pollen and spores this is particularly so, for the pollen rain at any site will be of complex origin and will be subject to both selective dispersion and filtration through growing vegetation (Tauber 1967). Although much of the pollen will be of local provenance, some will have been carried long distances and will come from vegetation formations a long way away. Because of these complications, microfossil evidence for the former presence or absence of a taxon at a given site can be difficult to use in plant geography. The pollen analysis of sediments is, however, the best available source of information for past regional and local environmental conditions.

Another problem is that pollen production varies from species to species (Andersen 1970, 1973), which means that low pollen producers, such as the insect-pollinated species of the maple genus *Acer,* will be little represented in the fossil record, whereas great pollen producers, like the wind-pollinated pine genus *Pinus,* will be over-represented. An additional difficulty is the selective decay of pollen taxa after deposition. The pollen of *Quercus* (oaks), for example, appears to be less resistant to weathering than that of *Pinus* (Havinga 1964). In order to understand all these problems more fully it is now common for studies to be made of the relationships between plant communities and the pollen rain being produced (e.g. Davis 1967, Flenley 1973).

The detailed procedures of pollen analysis have received full treatment in a number of publications and are not repeated here. The techniques employed in the Sub-department of Quaternary Research at the University of Cambridge are summarised by West (1977, 418–23) and very clear discussions of both the methods in current use and their limitations are provided by Brown (1960), Erdtman (1943), Faegri and Iversen (1974) and Moore and Webb (1978). Basically, pollen analysis is the process of identifying and counting the frequency of types of pollen grains and spores in a sediment. The procedure usually involves a number of stages, including the removal of the organic and inorganic sediment containing the pollen, the preparation of pollen mounts, normally with glycerine jelly or silicone oil, and the process of pollen counting. Magnifications of at least × 300 are needed for routine counting, but for more detailed work × 1000 is essential. Counting is accomplished by making traverses at standard intervals (usually not less than 1½ field diameters apart) across the whole width of the slide. Each grain encountered during the traverses is identified and recorded. Determination is frequently made by reference to a previously prepared collection of reference slides, built up by using pollen taken from known living species. The procedure for counting diatoms is similar to that used in pollen analysis and a method for their extraction and mounting is provided by Evans in West (1977, 424).

In Pleistocene studies the results of pollen analysis are invariably presented in the form of pollen diagrams (Fig. 6.3) comprising a series of curves or histograms. In relative pollen diagrams, these show the percentage frequency of the

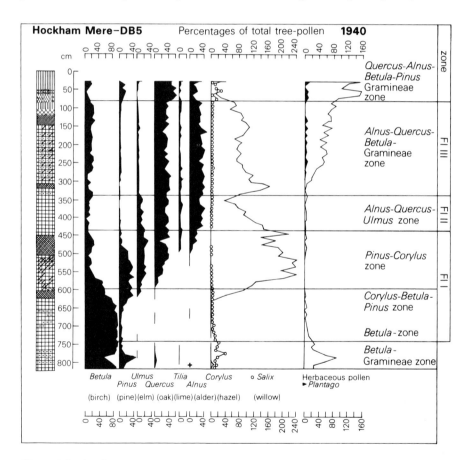

Figure 6.3 A pollen diagram through a lake deposit at Hockham Mere, Norfolk. The symbols in the column on the left refer to various types of limnic deposit up to 100 cm depth and to fen peat above this level (*after* Godwin and Tallantire 1951, Fig. 3; reproduced by permission of Blackwell Scientific Publications Ltd).

pollen and spore taxa identified at successive stratigraphic levels. An absolute diagram, on the other hand, shows the numbers of grains or spores deposited on unit surface area of sediment in unit time, or the numbers in unit volume of sediment.

The interpretation of pollen diagrams is a complex matter and is discussed at length by West (1977, 144–8). The main problem is to determine the exact relationship that existed between the pollen sedimentation at a site and the vegetation that produced it. Most pollen diagrams are zoned into a series of pollen assemblages, each one being characterised by a particular grouping of pollen taxa. This series of zones can give a reasonable idea of the vegetation sequences in the area from which the sediment received its pollen content, and the identification of such zones is one of the main ways of reconstructing the vegetation and climatic changes that occurred during the Pleistocene.

Microfossil evidence is also of increasing importance in the study of the earlier geological periods. In this instance, the microfossils are usually obtained by the

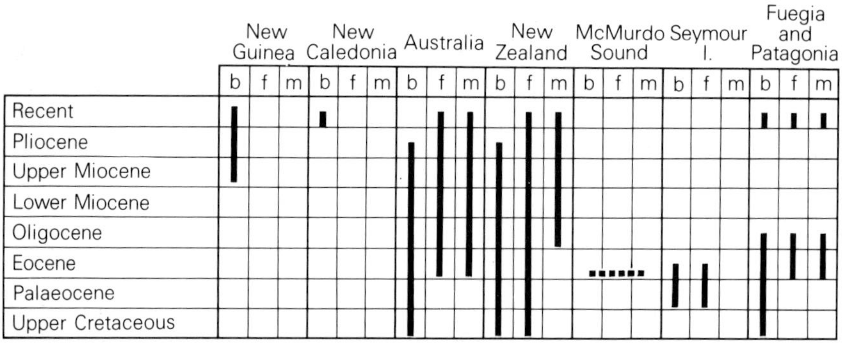

	New Guinea			New Caledonia			Australia			New Zealand			McMurdo Sound			Seymour I.			Fuegia and Patagonia		
	b	f	m	b	f	m	b	f	m	b	f	m	b	f	m	b	f	m	b	f	m
Recent																					
Pliocene																					
Upper Miocene																					
Lower Miocene																					
Oligocene																					
Eocene																					
Palaeocene																					
Upper Cretaceous																					

Figure 6.4   The stratigraphical record of *Nothofagus* (southern beech) pollen. The three pollen types are: b, *brassii* type; f, *fusca* type; m, *menziesii* type (*after* van Steenis 1972, Fig. 2; © Linnean Society of London).

chemical maceration of rock. Among the microfossils in Silurian and Devonian rocks, for example, there are many spores of vascular plants that can be described in detail, and to which generic and specific names can be applied. Such studies are providing important corroborative evidence to be considered along with that from the more generally used, but much fewer, macrofossil remains. As Banks (1970, 80) concludes, 'these microfossils probably represent a considerably larger sampling of the Devonian flora than do the macrofossils'.

In plant geography, the detailed study of the taxonomic value of microfossil remains, especially of pollen grains, is also providing important new lines of evidence by which to interpret the evolution and past distribution of certain taxa. The analysis of fossil pollen in the southern beech genus *Nothofagus* is a case in point. Figure 6.4 shows the stratigraphical record of *Nothofagus* pollen from the Cretaceous to the present (van Steenis 1971, 1972). It will be noticed that the fossil evidence is divided into three distinct pollen types that are named after three specific epithets but which in fact do not coincide with the normal taxonomic subdivisions of the genus. From the study of living plants, it appears that one 'pollen species' may be common to several taxonomic species and the development of the genus from the Cretaceous onwards may actually run to 100 taxonomic species. Certain gaps in the fossil record, such as the older Tertiary in New Guinea, are expected to be filled by future research.

Historical evidence

On cursory consideration, the historical evidence for the past distribution of plants found in written and published sources, such as Floras or expeditionary accounts, might be expected to prove less difficult to interpret than the enigmatic fossil record of the palaeobotanist. Unfortunately, this is not always the case and many written records raise more problems than they solve. The first, and most obvious, group of problems concern the identification of the plant in question.

Without a well curated and carefully labelled herbarium specimen to support a written record, the taxonomic status of a record may always be in doubt (see Ch. 2, pp. 13–14). The original recorder may have been certain of his identification, but how sure are we that he got it right? Moreover, with subsequent taxonomic and nomenclatural revisions it may prove impossible to know exactly which modern taxon is involved and, in certain extreme cases, the name and the identification may be untraceable in modern science. For these reasons, Rose and James (1974), in their definitive survey of the corticolous and lignicolous lichen species of the New Forest, Hampshire, distinguish carefully between records that can be verified by specimens in herbaria or by current field studies and those which are based on literature citations only. The acceptability of these literature citations is then assessed on ecological grounds and some are discarded as unlikely, such as the record for *Collema limosum* which is a species of base-rich soils and which would not be expected to grow on the acid soils of the New Forest.

But the problem is not only one of erroneous or difficult records. It is also related to the selective nature of recording itself. Sometimes the interest is not in what is recorded but in what has been missed or not recorded, and it is noteworthy that Rose and James include a category of lichens which might be expected to have occurred or to occur in the New Forest but which have never been recorded at all. If the value of historical records for establishing the former presence of a taxon is limited, their value for indicating absence is even more so.

This difficulty is particularly well illustrated by the 1888 record of Masclef for the presence of the lizard orchid (*Himantoglossum hircinum*), a distinctive orchid not likely to be misidentified, in the *département* of Pas-de-Calais, Northern France. He wrote that [the author's translation] 'the only locality actually known in the hills of Artois is that at Camblain-Châtelain, and even here it appears to be near extinction'. At face value, it would seem that this was the sole site for the plant at the time Masclef was writing, but many questions need to be asked and answered before we can arrive at a true understanding of the exact value of the record. How thorough was Masclef in his field botany? How many possible sites had he actually visited? How many likely sites existed? Could he have overlooked other sites? Was he in correspondence with botanists in other regions who might also have known the plant? Did he take account of older records? Today the plant is not uncommon in the *département* and we must therefore conclude that either Masclef was not very thorough or that it has spread to new sites since the 1880s. As it happens, the detailed work of Good (1936) on the geographical history of *H. hircinum* in Britain indicates that the Masclef record was correct, and that the plant has markedly increased its distribution during the twentieth century in response to the amelioration of winter temperatures which became notably accentuated after 1900. But without reference to this independent study the record as it stands would have proved very difficult to interpret.

The careful use of historical evidence therefore involves a number of important considerations, including the application of ecological and phytogeographical common sense, the study of the recorder as well as the record and the thorough search for supporting evidence wherever possible. Yet, even when such care is exercised, a large number of records will still remain obscure and uninterpretable. One frequent cause for disappointment is the vague nature of many geographical references which prove to be too general for any practical purposes. Rose and James, for example, have to acknowledge one group of lichens based on unlocalised specimens labelled 'New Forest' only.

The retrospective method in plant geography

In the best of all possible worlds the plant geographer would possess a full fossil and historical record for the past distributions of the plants that interest him. In reality, of course, the record is distinctly patchy, difficult to interpret and frequently non-existent. A wide range of taxa are rarely found fossil at all and many types of habitat yield few fossils. For example, the survival of pollen grains is much inhibited in calcareous habitats, and for wide stretches of chalk and limestone country there may be little chance of obtaining satisfactory evidence for past vegetation changes using pollen analysis. Moreover, only a limited number of species produce pollen which is diagnostic at the species level, and although pollen morphology is usually typical of a family or a genus it is often impossible to say with certainty whether the pollen of a rare species has been recorded or not. Of the present-day rarities listed by Pigott (1956) for the Upper Teesdale flora, only eight uncovered in Flandrian deposits have been found to possess pollen clearly diagnostic at the species level (Squires 1978). Another six have pollen identifiable at a level close enough to that of the species for one to be fairly confident that the pollen grains represent the rare plants concerned.

Because of these limitations, the plant geographer must often make recourse to other methods of determining the causes and origins of present distributions and the nature of past distribution patterns. In these methods, the present is seen as the key to the past and the detailed study of present patterns is used to generate theories concerning the history of distribution. Unfortunately, many of these theories remain unprovable and untestable and the aim is to arrive at the most elegant interpretation of what limited evidence exists.

The basis of this retrospective method is the identification of groups of taxa which possess markedly similar distribution patterns, and which appear on both ecological and phytogeographical grounds to have experienced a parallel history in fairly recent times. In practice this usually involves the recognition of taxa with distinctive disjunct or endemic distributions (see Chs 7 and 8) that can be reasonably explained in the light of our knowledge of past climatic and geomorphological changes. Theories are thus established to explain the present status and history of particular groups of taxa, taking into account the general evidence concerning the geological past, which has been derived from geological, geomorphological and palaeoecological studies.

The essential nature of the method is particularly well exemplified in the work of Pigott and Walters (1954) on the interpretation of the discontinuous distributions shown by certain British species of open habitats, which otherwise exhibit more widespread and more nearly continuous distributions over a large part of the European continent. It has long been recognised that many of these species, which belong to the 'continental' type of distribution or to the more southern and western Mediterranean elements, are not only rare but also tend to occur together in the same habitat, often in very restricted and disjunct areas. A careful study of these habitats revealed two factors in common between them: first, the presence of basic rock or base-rich drainage water, and secondly, the absence of naturally dominant woodland. Moreover, it was known from fossil plant remains that some of the species involved, such as the shrubby cinquefoil (*Potentilla fruticosa*), mountain avens (*Dryas octopetala*), Jacob's ladder (*Polemonium caeruleum*) and certain rockroses (*Helianthemum* spp.), had been an integral part of the open habitat 'park tundra' communities of the late-glacial period and they appear to have possessed much wider distributions at that time on the base-rich glacial soils.

Pigott and Walters therefore interpreted the present restricted and disjunct distributions of some of these species as late-glacial relict distributions which survived the spread of the Boreal birch–pine and Atlantic deciduous forests of the post-glacial period, on those sites which could never have carried closed woodland, even during the forest maximum. From our knowledge of the present-day vegetation, they identified the following types of habitat as possible refugia for these relicts:

(a) mountains above the tree limit;
(b) inland cliffs and screes;
(c) sea cliffs;

(d) river gorges, eroded river banks, river shingle or alluvium;
(e) sand dunes and dune 'slacks';
(f) shallow soils over chalk and limestone, especially on steep slopes;
(g) certain marsh and fen communities and lake shores.

As Pigott and Walters point out, it is significant that it is precisely in these types of habitat that we today find the local and rare plants that possess disjunct distributions. Of course, there is no one site at which all the members of this late-glacial relict group are present, although many are found in a wide range of the habitat types recognised as open habitat refugia. Moreover, the species belong to a number of different floristic elements, such as the arctic-alpine mountain avens (*Dryas octopetala*) and the continental purple milk-vetch (*Astragalus danicus*). The picture is not a simple one of all the taxa being of the same geographical element and occurring in exactly the same locations. Reality is much more complex, with some groups surviving at one site, another group elsewhere, and so on. The retrospective method, however, has been used to draw all the evidence together and to interpret this in the light of our knowledge of the late-glacial and post-glacial periods. The present distribution patterns of these species of open habitats have given the clear hint that their histories might have something in common, just as their present habitats are all characterised by base-rich soils and the absence of naturally dominant woodland.

In many cases more detailed investigations into the autecology, numerical taxonomy, chemotaxonomy and **cytology** of the plant populations involved in these studies can go a long way to help to elaborate the hypothesis put forward. Recent work by Fearn (1972), for example, on the cytology of the horseshoe vetch (*Hippocrepis comosa*), a frequent plant of dry calcareous grasslands in southern Britain and in central and southern Europe, has identified three cyto-types with very different phytogeographical histories:

  (i) a **hexaploid** (confined to a small area of the Pyrénées);
 (ii) a **tetraploid**;
(iii) a diploid.

In Britain both the diploid and the tetraploid races occur, and their known distribution is shown in Figure 6.5. The tetraploid is well distributed, whereas the diploid race possesses a disjunct distribution, growing mainly in south-western Britain, but also in Derbyshire. The sites for the diploid populations include a number of localities noted for their assemblages of rare relict species, and the diploid distribution parallels that of three other species very closely, namely the white rockrose (*Helianthemum apenninum*), the hoary rockrose (*H. canum*) and goldilocks (*Aster linosyris*). It seems that these three species and the diploid race of *H. comosa* are further examples of relicts from the late-glacial steppe–tundra flora as discussed by Pigott and Walters. The diploid race of *H. comosa* appears to be intolerant of competition and *H. comosa* as a species is intolerant of shade. It seems that these are the reasons why the diploid populations have persisted only on very sparsely vegetated, open habitats, which would not have been fully wooded at the time of the post-glacial forest maximum. The tetraploid race probably also survived on such sites, but because it is more tolerant of grazing and competition, it was able to recolonise the new tree-less base-rich habitats created by man and his grazing animals during the

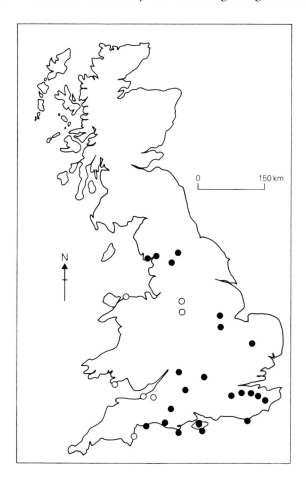

Figure 6.5 The known distribution of diploid (O) and tetraploid (●) populations of the horseshoe vetch (*Hippocrepis comosa*) in Britain (*after* Fearn 1972, Fig. 1).

Neolithic and Bronze Ages. The tetraploid race may also be more climatically tolerant of the changes in winter temperature northwards.

By such detailed studies of present-day plant populations, the plant geographer is able to propose reasonable and acceptable theories to account for their past and present distribution. It cannot be emphasised too much, however, that this retrospective method can never replace, or be as satisfactory as, the use of direct evidence based on fossil or historical records, and that, wherever possible, all sources of evidence should be employed together to arrive at the most elegant explanation.

Further reading

Useful introductory texts are marked with an asterisk.

General

*Banks, H. P. 1970. *Evolution and plants of the past*. Belmont, Calif.: Wadsworth.
Hallam, A. 1972. Continental drift and the fossil record. *Scient. Am.* **227,** 56–70.
Ross, C. A. (ed.) 1976. *Paleobiogeography*. Stroudsburg, Pennsylvania: Dowden, Hutchinson & Ross. (A collection of important papers spanning a century of publication on the interpretation of fossil distributions. The papers document in a very immediate way the debates on the permanency of continents and continental drift.)

Palaeobotanical evidence

*Andrews, H. N. 1961. *Studies in paleobotany*. New York: John Wiley.
Godwin, H. 1975. *The history of the British Flora. A factual basis for phytogeography,* 2nd edn. Cambridge: Cambridge University Press.
Kummel, B. and Raup, D. (eds) 1965. *Handbook of paleontological techniques*. San Francisco: W. H. Freeman.
*Moore, P. D. and J. A. Webb 1978. *An illustrated guide to pollen analysis*. London: Hodder & Stoughton.
West, R. G. 1977. *Pleistocene geology and botany,* 2nd edn. London: Longman.

The written record

Sheail, J. and T. C. E. Wells 1980. The Marchioness of Huntley: the written record and the herbarium. *Biol. J. Linn. Soc.* **13,** 315–30.

The retrospective method

Fearn, G. M. 1972. The distribution of intraspecific chromosome races of *Hippocrepis comosa* L. and their phytogeographical significance. *New Phytol.* **71,** 1221–5.
Holmquist, C. 1962. The relict concept. *Oikos* **13,** 262–92.
Pigott, C. D. and S. M. Walters 1954. On the interpretation of the discontinuous distributions shown by certain British species of open habitats. *J. Ecol.* **42,** 95–116. (A classic paper.)

7   Interpreting disjunct distribution patterns

We have already seen that there is probably no such thing in nature as an absolutely continuous distribution of a species, even over small geographical areas. All distributions are therefore to some extent broken or disjunct. However, in many cases the disjunction reflects no more than the normal dispersal capacity of the species concerned, or an easily explained ecological barrier. Such barriers may be topographic, climatic, **edaphic** or biological in character. If the disjunctions so caused are not great, the distribution pattern may still be regarded as continuous at the wider scale. For example, many aquatic species, such as the reed (*Phragmites communis*) and the hornwort *(Ceratophyllum demersum)* (Fig. 7.1), are considered cosmopolitan or sub-cosmopolitan at the world scale, even though they are absent from vast areas of dry land. *P. communis* is thought to be the most widely distributed of all flowering plant species.

The biogeographical term, disjunction, is therefore used in a much narrower sense and is applied to those distributions in which two or more populations of a taxon are exceptionally widely separated *for the taxon concerned.* Discontinuous or disjunct families are thus mainly considered at the world scale, although the

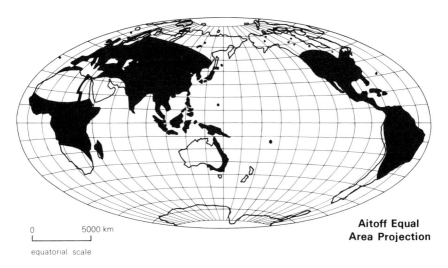

Figure 7.1 A much generalised distribution map of the Sub-cosmopolitan aquatic species *Ceratophyllum demersum* (the hornwort). Note the use of the Aitoff Equal-area projection and the bar scale for distances at the Equator (*after* Thorne 1972, Fig. 43).

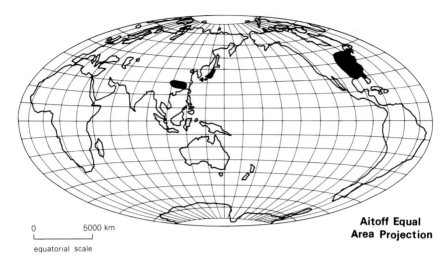

0   5000 km

equatorial scale

Aitoff Equal Area Projection

Figure 7.2   The generalised distribution of the Asian–Eastern American genus *Hamamelis,* the witch-hazels. Note the use of the Aitoff Equal-area projection and the bar scale for distances at the Equator (*after* Thorne 1972, Fig. 13).

disjunction of a species distribution may be studied at various levels, ranging from within a very small area, or a single country, or a continent, to the global level. Figure 7.2 shows the world distribution of the witch-hazel genus *(Hamamelis),* which today is found only in eastern North America and east Asia, a disjunct distribution pattern repeated by no less than some 74 different genera (Gray 1846, 1859, Li 1952, Thorne 1972, Wood 1971). It is with the interpretation of such major disjunct distribution patterns that this chapter is concerned.

A theoretical basis

There is no one explanation for the disjunct distribution patterns displayed by the world's plant taxa. In fact, a case can be argued for the distribution pattern of every taxon to be studied solely on its merits, with the expectation that each pattern will have something unique about it. It is indeed difficult to move from the form – the disjunct distribution – to an understanding of the processes which have brought about that form, for it is clear that the same form can be produced by widely differing processes (Fryxell 1967). The interpretation of disjunct distributions is therefore a complex and often impossible task.

However, in seeking a theoretical basis for the interpretation of disjunct distribution patterns, it is possible to identify a series of frequently recurring explanations. These are summarised graphically in Figure 7.3 for the simple case of a taxon, *X,* which possesses but two geographically isolated populations, x_1 and x_2. The scale involved in this disjunction need not worry us at present. In theory, the possible explanations could apply to the distribution of weeds on a lawn or to a major disjunct taxon at the world scale. In practice, they should be restricted to the study of biogeographical distributions displaying genuine disjunction in the narrower sense already defined. Each of the cases presented in Figure 7.3 will now be considered in turn. They are discussed as alternative explanations, even though in reality disjunctions are rarely monocausal and are often only explicable in terms of a combination of the cases outlined.

Case 1. In Case 1, it is assumed that the plants in population x_1 are taxonomically the same as those comprising population x_2, and that the disjunction therefore

Taxon **X** has two geographically isolated populations, **X₁** and **X₂**

The following represent the possible interpretations of this distribution pattern.

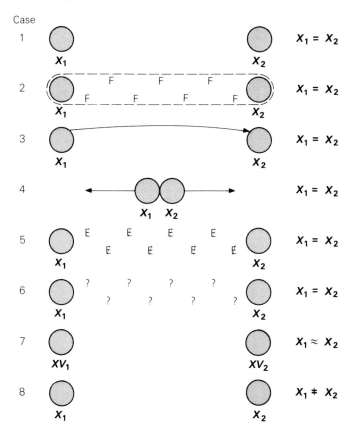

Figure 7.3 Interpreting disjunct distribution patterns. For a full explanation see the text, pp. 92–105.

appears to be genuine. The explanation for this disjunction is that the same taxon has evolved twice, quite independently, but in two widely differing geographical localities. So far the two populations have not enlarged their geographical range to any significant extent and therefore remain disjunct.

Species so arising are termed polytopic and the process by which they evolve polytopy. If the separate populations have developed at different times, they are also polychronic in origin. At the present, such an explanation for a disjunction would be accepted only in exceptional circumstances. Although the same

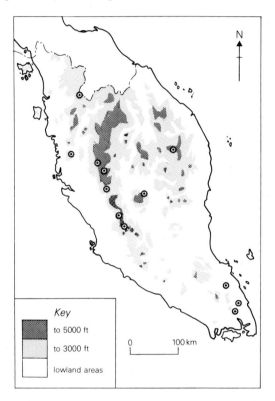

Figure 7.4   The distribution of *Macaranga quadricornis* in peninsular Malaysia. Note that it has two geographically and ecologically disjunct areas (*after* Whitmore 1969, Fig. 5; © Linnean Society of London).

mutations have been repeated in populations under experimental conditions, the odds against this occurring in nature seem very long with the intense competition and extreme diversity that characterise natural habitats. As Polunin (1960, 207) comments, 'it scarcely seems conceivable for members of major species whose common ancestry was remote and whose distinctive characters are numerous'.

Nevertheless, Whitmore (1969) has invoked this explanation to account for the disjunct distribution of a more 'minor' species, *Macaranga quadricornis*, in the Malay peninsula. *M. quadricornis* is endemic to Malaya, where it is one of the most beautiful and conspicuous wayside trees of the main mountains of the peninsula, south to Gunong Nuang at 3°20′N (Fig. 7.4). Whitmore says that 'despite thorough searching it has not been found lower than 890 m, nor has it

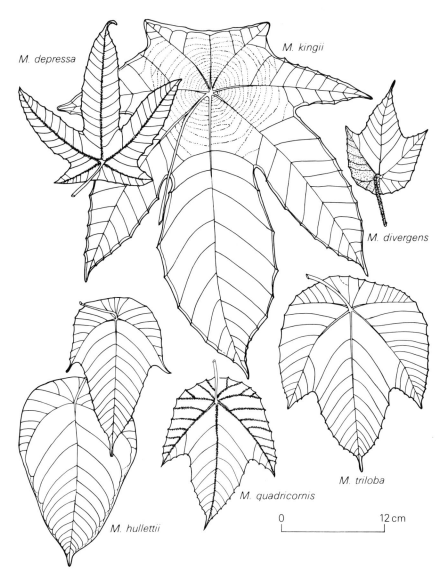

Figure 7.5 The leaf forms of *Macaranga triloba* and its allies (section *Pachystemon*), including *M. quadricornis* (*after* Whitmore 1969, Fig. 3 – the original drawing by Yusoff bin Haji Mohd. Saman; © Linnean Society of London).

been found on the isolated mountain ranges of south Malaya'. However, the same taxon re-occurs in the lowland and seasonal swamps of Johore, north to 2°20'N. The only differences Whitmore could detect in these plants were a slightly longer central lobe to the leaf of juvenile trees and 'an indefinable difference in the velvety texture of the living leaves, which disappears on drying'.

The species therefore consists of two disjunct groups which are widely separated in altitude, geographical range and habitat. In coming to a conclusion on the origins of this disjunction, Whitmore notes that *M. quadricornis* is in fact a member of a close-knit complex of species, which he terms the *M. triloba* group (Fig. 7.5), and which has representatives widely distributed throughout western Malesia. He argues that *M. quadricornis* has evolved twice from out of this complex and in two separate localities with markedly different habitats. Single chromosome counts from *M. triloba*, *M. quadricornis* and another species of the group were all $2n = 22$. *M. quadricornis* could, therefore, be an example of polytopy not involving polyploidy (see Chapter 9).

Further possible cases of polytopic evolution, this time involving polyploidy, are discussed in Davis and Heywood (1963).

Case 2. As in Case 1, the disjunction is again accepted as genuine, the individuals of the two populations clearly representing the same taxon. In this case, however, there is acceptable historical or palaeobotanical evidence that in the geological past the taxon was much more widely distributed than at present, and the now separated populations were once linked by a more continuous distribution. The present disjunction is therefore a relict one and X is a relict taxon

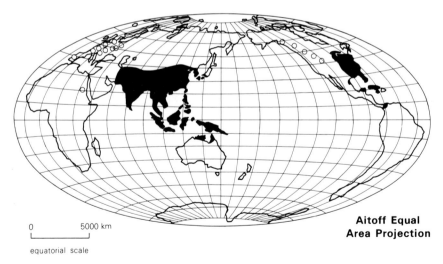

0 ____ 5000 km

equatorial scale

**Aitoff Equal
Area Projection**

Figure 7.6 The past and present distribution of the water-lotus genus *Nelumbo*. The open circles represent fossil records. Note the use of the Aitoff Equal-area projection and the bar scale for distances at the Equator (*after* Thorne 1972, Fig. 14).

which has survived in its two separate localities even though the surrounding areas have been vacated. The sites at which it has persisted are termed refuges or refugia, and they probably possess some distinctive ecological, geomorphological or geological characteristics which account for the survival there of the relict taxon. In the absence of satisfactory historical and palaeobotanical evidence, the identification of such refugia may help to establish a relict theory of disjunction, as discussed at length in the previous chapter.

A good example of a relict taxon now exhibiting a markedly disjunct distribution is given in Figure 7.6, which shows the past and present distribution of the Asian–Eastern American water-lotus genus, *Nelumbo*. The fossil record for this genus clearly indicates that, during the Tertiary, it possessed a nearly continuous distribution across Eurasia and North America. The same story also holds for a range of other genera, such as the tulip-tree genus *(Liriodendron)* and *Magnolia* (Meusel 1969), which today display a similar Asia–Eastern American disjunction.

Liquidambar
Hamamelis

Also
ginseng

Also
Limulus

Case 3. Once more, the disjunction is regarded as real, with population x_1 representing the same taxon as x_2. This time, however, the disjunction is explained in terms of exceptional long-range dispersal, taxon X having dispersed itself from its original population x_1, across intervening barriers, be they oceans or mountains, to establish the second population x_2, and thus create a disjunct distribution.

This efficacy and likelihood of such long-range dispersal in seed plants remain subjects of great controversy. For a new population of plants to become permanently developed three main phases are necessary (Solbrig 1972). These are (1) effective dispersal, (2) germination and establishment and (3) survival through time. Even if a plant propagule is transported over long distances, there is no guarantee that it will be able to grow and live at the new locality. As Solbrig writes, 'the dispersal capacity of any propagule is always greater than its capacity to germinate and to become established at the site to which it was dispersed, and likewise, a plant is capable of germinating and growing in more places than those where it is capable of surviving generation after generation'.

This very same point was made by Darwin in *The origin of species*. 'It should be observed', he wrote, 'that scarcely any means of transport would carry seeds for very great distances; for seeds do not retain their vitality when exposed for great lengths of time to the action of sea-water . . . Ocean currents, from their course, would never bring seeds from North America to Britain [see Fig. 7.7], though they might and do bring seeds from the West Indies to our western shores, where if not killed by so long immersion in salt-water, they could not endure our climate.' In a recent study, Nelson (1978) has recorded the tropical drift fruits and seeds which have made their way to Irish beaches. These indeed prove to be mainly of West Indian origin and their recovery sites are shown in Figure 7.8. Although these tropical plants are not likely to become established in Ireland, Nelson's study clearly shows that their disseminules must, at least, be capable of travelling in sea for over one year.

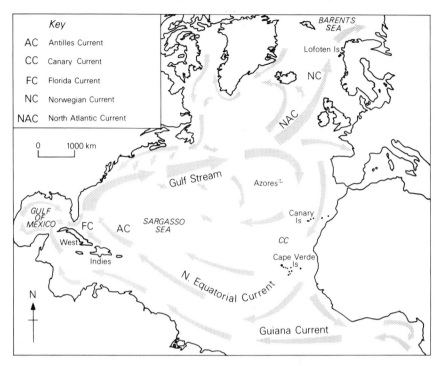

Figure 7.7   A generalised map of the surface currents in the North Atlantic Ocean (*after* Nelson 1978, Fig. 2; reproduced by kind permission of the author).

The possibility of exceptional long-range dispersal is thus indicated for plants with fruits able to float and remain viable for long periods in salt water (Guppy 1906, Gunn & Dennis 1976). It is well known that the coconut *(Cocos nucifera)*, for example, can remain unharmed in sea water for a length of time permitting it to drift no less than 3000 miles. Moreover, Sykes and Godley (1968) have produced convincing evidence for the transoceanic dispersal of *Sophora microphylla* from New Zealand to south Chile and Gough Island, a dispersal direction well in accordance with the easterly orientation of the winds and ocean currents in that part of the southern hemisphere.

Along with ocean currents, the most probable agencies of long-range dispersal are wind and animals, especially migrating birds (Ridley 1930). However, the number of plants displaying adaptations for such modes of transport tends to be limited. In a study of the 183 montane species of the Loma mountains and Tingi hills of West Africa, Morton (1972), for example, found only 40 (22%) which possessed seed characters likely to lead to long-range dispersal by wind or birds (see Table 8.2, Ch. 8). Yet it is important to remember in all discussions concerning the efficacy of long-distance dispersal that the event *needs to succeed only once* (though more often if the plant is **dioecious**) and a new population will be established. We are thus primarily concerned with the identification of exceptional occurrences and not the analysis

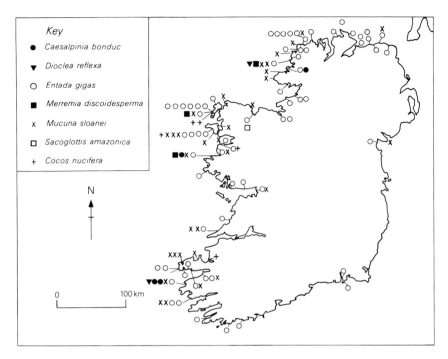

Key

● Caesalpinia bonduc

▼ Dioclea reflexa

○ Entada gigas

■ Merremia discoidesperma

x Mucuna sloanei

□ Sacoglottis amazonica

+ Cocos nucifera

N

0   100 km

Figure 7.8 The recovery sites of peregrine drift fruits and seeds in Ireland. Each symbol indicates a single specimen (*after* Nelson 1978, Fig 1; reproduced by kind permission of the author).

of a steady process. It is this fact which makes the study of long-distance dispersal so particularly difficult.

Shorter range dispersals, from mountain to mountain, island to island or across narrow water barriers, are, however, of unquestionable significance in creating new distribution patterns, and over a long period of time a series of small steps may well lead to a pattern which is in the end widely disjunct. Thorne (1972) thus argues that short-distance dispersal provides an adequate explanation for the current disjunction of many genera found in Asia, Papua and Australasia. He envisages that these distributions were formed during periods, such as the glacial lowering of sea levels, when much more land connected Malesia to mainland Asia and Australia. The possible importance of such 'land bridges' has been discussed at length by van Steenis (1962).

Case 4. Again the present disjunct distribution of taxon *X* is regarded as genuine. In this case, however, it is argued that the now disjunct populations were originally one and have, at some time in the geological past, been divided and have moved apart. Obviously, the most important example of such a case would be the separation of populations through the agency of continental drift, plate tectonics and ocean-floor spreading, subjects which have been discussed in detail in Chapter 5. Although this explanation remains controversial, with the

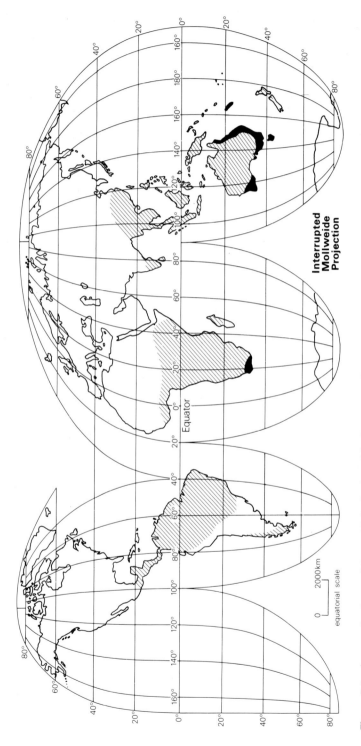

Figure 7.9 A generalised map of the world distribution of the Proteaceae. Regions of major generic diversity are shown in solid black. Note the use of the Interrupted Mollweide Projection (*after* Johnson and Briggs 1975, Fig 4; © Linnean Society of London).

establishment of plate tectonics as a fundamental principle of earth science possible past continental movements have been increasingly in vogue with plant geographers as a theory to account for disjunct distribution patterns.

Indeed, Raven and Axelrod (1974) have put forward a comprehensive, if naturally speculative, series of explanations for flowering plant distribution in the light of such past continental movements. The chronology they establish has been employed by Johnson and Briggs (1975) to account for the relevant separations and migrations of taxa in the important southern family, Proteaceae (Fig. 7.9). Their complex analysis is summarised in Table 7.1. Johnson and Briggs wisely emphasise that their conclusions are clearly dependent on (a) long-distance dispersal being ruled out in many instances, (b) the interpretations of plate tectonics proving accurate and (c) their **phylogenetic** system being correct. If these assumptions indeed prove to be firmly founded, they will have provided a satisfactory explanation, in terms of plate tectonic events, for such classic proteaceous disjunctions as the Proteoideae (Proteeae – Africa; other Proteoideae – Australasia) and *Oreocallis* (South America – Australasia and New Guinea).

Cases 5 and 6. Only a brief mention of these two cases is necessary, although they can be of great importance in the interpretation of disjunct distributions. In both instances the plants in population x_1 are once more taken to be the same as those comprising population x_2.

In case 5, the explanation for the disjunction is ecological, populations x_1 and x_2 being developed in the only two localities with habitat conditions suitable for the germination, establishment and growth of taxon X. Obviously habitat disjunction is a significant explanatory factor in many short-range disjunctions and especially for taxa with highly specialised habitat requirements e.g. aquatics, and species adapted to soils rich in heavy metals.

In case 6, however, the disjunction may not in fact be real, for between the two known populations of taxon X, there are probably further, linking populations as yet undiscovered and unrecorded. When discussing disjunct distribution patterns, this possibility should always be kept in mind, especially if analysing distributions for those parts of the world which remain poorly studied.

Case 7. This case is a distinctive and very important variant of the disjunct distribution. In contrast to the preceding six cases, the taxa comprising populations x_1 and x_2 are not exactly the same. They are, however, very close to each other taxonomically, and represent a species-pair (Xv_1 and Xv_2) which have probably evolved from a common ancestor.

Such species are known as vicarious species. They are closely related **allopatric** species which have descended from a common ancestral population and which have attained spatial isolation, hence their disjunct distribution. Whenever plant populations become separated from each other, they are subject to different selection pressures, and such a state is the first step towards differentiation and possible eventual speciation (Mayr 1963, Solbrig 1972,

Table 7.1 Plate tectonic events and the possible distributional history of the Proteaceae. Based on the chronology of Raven and Axelrod (1974) and on the assumptions outlined on this page (after Johnson & Briggs 1975, Table 6; © Linnean Society of London).

Key

|| = Severance of land surfaces

⇌ = Conjunction of land surfaces

// = Presumed spatial and evolutionary separation of related taxa

→ = Presumed direction of spread following establishment of opportunity

[ ] = Hypothetical or fossil occurrences not represented by present-day taxa

'Aust.' = Australia

Note: probable source groups are given on the right in the final column.

Approx. time of changed migration opportunity* (m.y. BP)	Significant tectonic events	Relevant separations and migrations of taxa	
4 late Pliocene	orogeny in New Guinea	Grevilleeae	*Grevillea* spp. N. G. ← *Grevillea* spp. Aust.
		Banksiinae	*Banksia* sp. N. G. ← *Banksia* spp. Aust.
6 late Miocene	N. Amer. ⇌ S. Amer.	Roupalinae	*Roupala* spp. Central Amer. ← *Roupala* spp. S. Amer.
		Macadamiinae	*Panopsis* spp. Central Amer. ← *Panopsis* spp. S. Amer.
15 Miocene	Asia ⇌ New Guinea & Australia	Stenocarpinae	*Stenocarpus* spp. N. G. ← *Stenocarpus* spp. Aust.
		Embothriinae	*Oreocallis* sp. N. G. ← *Oreocallis* spp. Aust.
		Grevilleeae	*Finschia* N. G. & Melanesia ← *Grevillea* spp. Aust.
		Grevilleeae	*Grevillea* sp. Celebes ← *Grevillea* spp. Aust.
		Heliciinae	*Helicia* spp. Asia ← *Helicia* spp. N. G. ← *Helicia* spp. Aust.
		Hicksbeachiinae	*Heliciopsis* S. E. Asia ← other Hicksbeachiinae Aust.
		Macadamiinae	*Macadamia* sp. Celebes ← *Macadamia* spp. Aust.
17 Miocene	Africa ⇌ Eurasia	—	—
35 Oligocene	S. Amer. || Antarctica || Aust. (end of temperate link)	Oriteae	*Orites* spp. S. Amer. // *Orites* spp. Aust.
		Lomatiinae	*Lomatia* spp. S. Amer. // *Lomatia* spp. Aust.
		Embothriinae	*Oreocallis* spp. S. Amer. // *Oreocallis* spp. Aust. & N. G.
		Embothriinae	*Embothrium* S. Amer. // other Embothriinae Aust.
		Gevuininae	*Gevuina* sp. S. Amer. // *Gevuina* spp. Australasia

45	Eocene	Asia ⇌ India	Hicksbeachiinae	Heliciopsis S. E. Asia ← [Heliciopsis India]
63	early Palaeocene	Eurasia ‖ Africa	—	—
80	Senonian	N. Cal. ‖ N. Z. ‖ Australia	Persooniinae	Garnieria N. Cal. // Toronia N.Z. // other Persooniinae Aust.
			Cenarrheninae	Beauprea & Beaupreopsis N. Cal. // other Cenarrheninae Aust.
			Knightiinae	Eucarpha N. Cal. // Knightia N.Z. // Darlingia Aust.
			Stenocarpinae	Stenocarpus spp. N. Cal. // Stenocarpus spp. Aust. & N. G.
			Grevilleeae	Grevillea spp. N. Cal. // Grevillea spp. Aust.
			Gevuininae	Sleumerodendron N. Cal. // Gevuina spp. Aust. & N. G. (& S. Amer.)
			Gevuininae	Gevuina spp. Melanesia // Gevuina spp. Aust. & N. G.
			Hicksbeachiinae	Virotia spp. N. Cal. // Virotia sp.? Aust.
			Banksieae	[Banksieae N.Z. // Banksieae Aust.]
90	Turonian	Africa ‖ S. America	Macadamiinae	Brabeium Africa // Panopsis S. Amer.
100	Cenomanian	India ‖ Madagascar ‖ Africa	Hicksbeachiinae	[Heliciopsis India] // Malagasia Madagascar
110±10	mid-Cretaceous	W. Gondwanaland (Africa & S. Amer.) ‖ Australasia (end warm-temperate or subtropical link)	PROTEOIDEAE	Proteeae Africa // other Proteoideae Australasia
			Conospermeae	Dilobeiinae Madagascar // Cenarrheninae Australasia
			Gevuininae	Euplassa S. Amer. // other Gevuininae Australasia
			Hicksbeachiinae	Malagasia–Heliciopsis line W. Gondwanaland // other Hicksbeachiinae Australasia
			Macadamiinae	Brabeium–Panopsis line W. Gondwanaland // Macadamia Australasia
			Roupalinae	Roupala S. Amer. // Kermadecia Australasia

*Dates given are those considered by Raven and Axelrod (1974) as significant for plant migration, allowing for some short-distance dispersal. Thus the South American landmasses finally separated about 100 m.y.BP but short-distance migration would appear to have been possible until about 90 m.y.BP. Similarly the South America–Antarctica–Australia severance is dated as more recent than the beginning of separation of the continents.

Stebbins 1950). Thus, if taxon X were restricted to two geographically isolated populations, then through time new taxa might arise in the different populations, creating a closely related and disjunct pair of taxa, the vicariants Xv_1 and Xv_2.

Stebbins and Day (1967) have made a detailed study of such a species-pair belonging to Thorne's (1972) Mediterranean–American category of intercontinental disjunctions. In fact they identify no less than 21 genera with vicarious related species in arid and semi-arid regions of the Old World and North America. The species-pair with which they are primarily concerned is *Plantago ovata – P. insularis*. *P. ovata* is found in the warmer, drier parts of the Mediterranean, the Canary Islands and across North Africa, eastwards to India (Fig. 7.10). *P. insularis,* on the other hand, comes from the deserts of south-east California, south-west Arizona, the north of Baja California and the coast and Channel Islands of southern California.

This pair of species, belonging to the section *Leucopsyllium,* are very close taxonomically. They resemble each other in their annual habit of growth, in their leaf and spike morphology, and in most features of floral morphology, seed shape and seed structure. They both possess the gametic chromosome number $n = 4$ and they are easily hybridised, their F_1 hybrid proving vigorous and partly fertile. Moreover, they are both characteristic of arid and semi-arid regions with warm, temperate climates.

Stebbins and Day argue that these represent vicarious forms which have evolved, very slowly, from a common ancestor, which probably spread from the Old World to the New by means of a Bering Strait 'land bridge' some time not more recent than the beginning of the Miocene, around 20 million years ago. At this period, and earlier, the climate of the Bering Strait region would probably have been warm enough to support the ancestor of the two plants. Today the shortest distance between the two species is 8000 miles (12 875 km), from north-west India to California.

Cain (1944) identifies four main types of vicarious distribution. The first of these is horizontal or geographic vicariance, of the type exemplified by *P. ovata* and *P. insularis*. The second type is altitudinal vicariance i.e. lowland/highland species-pairs. The third is habitat or ecological vicariance, as illustrated by the sea arrow-grass *(Triglochin maritima)* and the marsh arrow-grass *(T. palustris),* which occupy respectively salt-water and fresh-water marshes. The fourth is seasonal separation, a category originally defined by Wettstein (1908).

Case 8. The final case can be easily stated. In this instance, detailed taxonomic research reveals that the so-called disjunct taxon X actually comprises two populations which are not the same and which represent totally different taxa. The disjunction is thus proved to be false. This is a not too uncommon occurrence. For example, the listing of the otherwise all-American genus *Miconia* of the Melastomataceae in *Flora of West Tropical Africa* (Hutchinson & Dalziel 1954–68) has been identified as an error by Wurdack (1970). *Miconia africana* is in fact a *Leandra,* probably collected in Brazil! As Thorne (1972) comments, 'a mix-up in labels seems indicated'.

Guiding principles in the study of major disjunctions

There have been a number of attempts to classify the major disjunctions in the geographic range of seed plants (see, for example, Good 1974, 472–78). A well-argued and recent classification of intercontinental disjunctions is that of Thorne (1972), which is summarised here in Table 7.2. For each category of disjunction, an example of a characteristic taxon displaying such a distribution pattern is indicated. Many of these distributions are actually mapped in Thorne's paper, but additional maps of intercontinental disjunct distribution patterns may be found in the works of van Balgooy (1966), Florin (1963), Hultén (1968), Meusel *et al.* (1965), Meusel and Schubert (1971) and Wood (1971), among others.

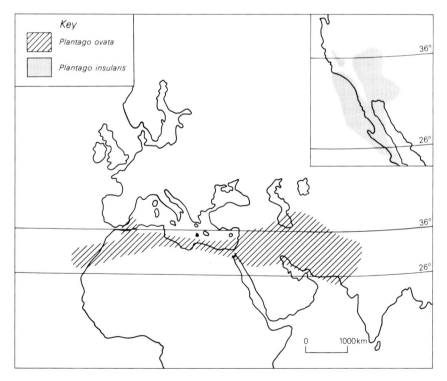

Figure 7.10 The geographical distribution of *Plantago ovata* (Canary Islands to India) and *P. insularis* (*inset,* south-west North America) (*after* Stebbins and Day 1967, Fig. 1).

Table 7.2   A classification of major disjunctions in the geographical ranges of seed plants (*after* Thorne 1972, 367).

I. Eurasian–North American
   1. Arctic
      1a. Circum-Arctic, e.g. *Diapensia lapponica*
      1b. Beringian–Arctic, e.g. *Aster sibiricus*
      1c. Amphi-Atlantic–Arctic, e.g. *Salix herbacea*

   2. Boreal
      2a. Circum-Boreal, e.g. *Scheuchzeria palustris*
      2b. Beringian–Boreal, e.g. *Oplopanax*
      2c. Amphi-Atlantic–Boreal, e.g. *Rhynchospora fusca*

   3. Temperate
      3a. Circum-North Temperate, e.g. *Pinus*
      3b. North and South Temperate, e.g. *Ribes*
      3c. Fragmentary North Temperate
         3c–1. Amphi-Atlantic Temperate, e.g. *Corema*
         3c–2. Mediterranean–American, e.g. Resedaceae
         3c–3. Eurasian–Eastern and Western American, e.g. *Juglans*
         3c–4. Eurasian–Eastern American–Mexican, e.g. *Carpinus*
         3c–5. Asian–Eastern and Western American, e.g. *Trillium*
         3c–6. Eurasian–Eastern American, e.g. *Castanea*
         3c–7. Asian–Eastern American, e.g. *Hamamelis* (Fig 7.2)
         3c–8. Eurasian–Western American, e.g. *Paeonia*
         3c–9. Asian–Mexican Highland, e.g. *Mitrastemon*
      3d. Wide Intercontinental Disjuncts and Epibiotics, e.g. *Camassia, Viscum album*

 II. Amphi-Pacific Tropical, e.g. *Perrottetia*

 III. Pantropical, e.g. *Diospyros* (Fig 3.8)

 IV. African–Eurasian (–Pacific)
   1. African–Mediterranean, e.g. *Cytinus*
   2. African–Eurasian, e.g. *Trapa*
   3. African–Eurasian–Malesian, e.g. *Irvingia,* Pandaceae
   4. African–Eurasian–Pacific, e.g. *Bruguiera*
   5. African–Eurasian–Australasian, e.g. *Sonneratia*
   6. Indian Ocean–Eurasian (–Pacific), e.g. *Nepenthes*

 V. Amphi-Indian Ocean, e.g. *Hibbertia*

 VI. Asian–Pacific
   1. Asian–Papuan, e.g. *Hopea*
   2. Asian–Papuan–Melanesian, e.g. *Ailanthus*
   3. Asian–Papuan–Pacific Basin, e.g. *Fagraea*
   4. Asian–Papuan–Australasian, e.g. *Melaleuca, Zingiber*

 VII. Pacific Ocean, e.g. *Coprosma*

 VIII. Pacific–Indian–Atlantic Oceans, e.g. *Zostera, Thalassia*

IX. American–African, e.g. Turneraceae

X. North American–South American, e.g. Sarraceniaceae

XI. South American–Australasian
1. South American–Australasian e.g. *Nothofagus*
2. South American–Australasian–Asian, e.g. *Coriaria*
3. South American–Australasian–Madagascan, e.g. Winteraceae

XII. Temperate South American–Asian, e.g. Lardizabalaceae

XIII. Circum-South Temperate, e.g. *Acaena*

XIV. Circum-Antarctic, e.g. *Ranunculus biternatus*

[XV. Subcosmopolitan, e.g. *Ceratophyllum demersum* (Fig 7.1)]*

XVI. Anomalous, e.g. *Pilostyles*

*Not strictly a disjunct distribution as defined in Chapter 7.

In compiling his particular classification, Thorne was guided by a number of basic principles which should govern the study of all disjunct distribution patterns. These principles are as follows:

(a) only taxa that have been reliably revised should be seriously considered;
(b) only accurate distributional data should be used in the compilation of distribution maps;
(c) because almost all taxa have discontinuous ranges, only those taxa exhibiting distributional gaps greater than their normal dispersal capacity should be said to have a disjunct range;
(d) the present distribution of a taxon is not necessarily a guide to its past distribution;
(e) coincidence in range distribution, although certainly suggestive, does not mean that two taxa have similar dispersal histories – the same form can be produced by different processes;
(f) the lower the rank of a taxon, the more instructive is its disjunct range.

To these six important guidelines proposed by Thorne, it is possible to add a seventh, namely that:

(g) disjunct distribution patterns are rarely monocausal, being the product, in the main, of a number of different factors of distribution.

Further reading

Fryxell, P. A. 1967. The interpretation of disjunct distributions. *Taxon* **16**, 316–24.
Good, R. 1974. *The geography of the flowering plants*, 4th edn. London: Longman (Chs 4, 6, 11 & 12; also Part 2.)

Gunn, C. R. and J. V. Dennis 1976. *World guide to tropical drift seeds and fruits*. New York: Demeter Press.

Johnson, L. A. S. and B. G. Briggs 1975. On the Proteaceae – the evolution and classification of a southern family. *Bot. J. Linn. Soc.* **70,** 83–182.

Raven, P. H. and D. I. Axelrod 1974. Angiosperm biogeography and past continental movements. *Ann. Missouri Bot. Garden* **61,** 539–673.

Ridley, H. N. 1930. *The dispersal of plants throughout the world*. Ashford: L. Reeve.

Solbrig, O. T. 1972. New approaches to the study of disjunctions, with special emphasis on the American Amphitropical Desert Disjunctions. In *Taxonomy, phytogeography and evolution*, D. H. Valentine (ed.) 85–100. London & New York: Academic Press.

Steenis, C. G. G. J. van 1962. The land-bridge theory in botany. *Blumea* **11,** 235–542.

Thorne, R. F. 1972. Major disjunctions in the geographic ranges of seed plants. *Q. Rev. Biol.* **47,** 365–411.

8 Interpreting endemic distribution patterns

Without doubt, endemic taxa have a fascination for the plant geographer. They are usually very rare plants with narrow, but distinctive, distribution patterns. They also represent the geographical element which most naturally characterises the floristic uniqueness of a particular country or region. It is not surprising therefore that their distribution patterns are often thought to contain a great deal of phytogeographical information which careful analysis will reveal and which will be of great value in interpreting the phytogeographical history of different regions and floras.

Their interpretation is, however, far from straightforward and the concept of endemism remains an enigmatic one, a precise scientific definition of 'endemic' proving elusive. It is the purpose of this chapter to consider these problems and to review some of the principles which have been developed in the study of endemic distribution patterns. As with the disjunct patterns discussed in Chapter 7, it will be seen that no one interpretation or explanation will suffice for all cases and that it is necessary to classify endemics into a number of different types.

What is an endemic?

In theory, the concept of peculiarity or restriction to one limited region can be applied at any scale, whether it be the distribution of weeds on a single patch of lawn, of mosses between two bricks in a wall or of a taxon with a very narrow distribution when viewed from the world scale. In practice, however, such an all-embracing approach is of limited value and in plant geography at least the concept needs further qualification. Tradition has therefore circumscribed the range of the word 'endemic' so that in normal circumstances it is applied only to taxa with comparatively or abnormally restricted distributions at the world, continental or regional scales. An endemic distribution is thus one in which the given taxon is contained entirely within one continent, country or natural area. In this sense, the white spruce (*Picea glauca*) may be said to be endemic to North America, the coast redwood (*Sequoia sempervirens*) to coastal California and Oregon, and the genus *Dasynotus* to a few square kilometres in Idaho (Daubenmire 1978). Obviously, species endemism is common at all scales of enquiry, but in the case of the higher taxonomic ranks endemism is usually restricted to the wider world and continental levels. For example, continental endemism is of importance in the discussion of plant families, the average distribution of families being wider than one continent.

Even within this restricted view of endemism, however, the range of geographical distribution embraced is vast, ranging from taxa confined to only one or two localities in the world to taxa restricted in their distribution to one continent. In consequence, many writers differentiate between broad endemic ranges and narrow endemic ranges (the so-called 'local' or 'narrow' endemics). The latter represent taxa which are confined in their distribution to one small island, one mountain range or just a few restricted localities. The use of the term 'endemic' therefore has no hard and fast rules and the concept of endemism must be employed with scientific sensitivity and care.

Yet, if the definition of the basic concept is difficult, the interpretation of the origins of endemic distributions is even more so. Once again we see in plant geography that the same distributional form does not always indicate the same process. Endemics have therefore been long classified into various types, according to their presumed origins.

New or old?

The classification of endemic distribution types is complex, and its history and terminology have been reviewed recently by Favarger and Contandriopoulos (1961), Prentice (1976), Richardson (1978) and Stebbins and Major (1965). Just as with disjunct distributions, it is possible to identify a whole series of frequently recurring explanations, but in many ways these are not so clear cut and they basically represent subtle variations on the two extreme alternatives indicated in Figure 8.1, namely those of neoendemism and palaeoendemism.

These two main types of endemic distribution were distinguished as early as 1882 by Engler and have dominated work on endemism ever since. The differences between them are simply stated. Neoendemics, which are also variously called autochthonous, progressive or secondary endemics, represent 'new' taxa which have arisen by differential evolution in a particular area from which they have not yet spread or are unable to spread. Where the taxa concerned are critical 'species' or are regarded as below the rank of species, being subspecies or varieties, they are frequently termed microendemics. Palaeoendemics, on the other hand, are taxa which once possessed much wider distributions but which are now confined to a very limited portion of their former territory. They are therefore 'old' taxa, though in fact they need not be any older than some neoendemics. They do, however, have relict distributions in only one continent, country, region or locality. In many ways they are nothing more than the ultimate product of disjunction, the reduction of a formerly wide range to one final area. If a species has only two remaining sites and one is destroyed, it becomes an endemic. Palaeoendemics are variously termed conservative endemics, relict endemics or epibiotics.

The contrast between neoendemism and palaeoendemism is, in part, linked to the differences in the rates of evolution among organisms and the nature of the ecological challenges in different areas. Obviously, the creation of neoendemics will be favoured by rapid rates of evolution, whereas palaeoendemism will result where rates are slow in relation to the external pressures of climatic and geological change. Moreover, the nature of a plant distribution will alter through time and all species must, in effect, start as neoendemics and end as palaeoendemics (Richardson 1978). Between these extremes many species will lose their endemic status and will occupy large areas. Other species, however, may always remain basically endemic, their distribution being naturally

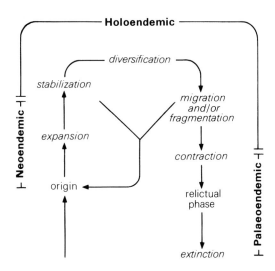

Figure 8.1 The evolution of taxa in a **monophyletic** assemblage, showing changes in geographical area and endemic status (*after* Richardson 1978, Fig. 1). (Reproduced with permission from H. E. Street (ed.) 1978. *Essays in plant taxonomy.* © Academic Press Inc. (London) Ltd.)

restricted by ecological and physiological constraints. In Figure 8.1, these are described as holoendemics and they are intermediate in character between the two extreme evolutionary forms.

In practice the recognition of these three categories of neo-, holo- and palaeo-endemic is far from easy. Richardson (1978) lists eight commonly occurring attributes which may help in the process of recognition, but there are many exceptions and in fact a continuum exists between the three types. The analysis of attributes is shown in Table 8.1. Only a combination of such attributes will provide a clear indication of the most likely endemic status for a given taxon.

As an example of neoendemism, the hawthorn genus (*Crataegus*) in south-eastern North America is interesting (Daubenmire 1978). These small trees are intolerant of shade, and before the advent of European immigrants there were probably few species, widely scattered. With the clearing of the forest and the creation of a large range of new habitats including forest margins, fence-rows, hedges and abandoned fields, the original taxa appear to have spread rapidly, being bird–disseminated and also being thorny plants able to withstand human pressure. Where formerly isolated populations and taxa came into contact, **hybrids** formed and gave rise to the present difficult taxonomic situation, with over 1100 species and varieties having been described, many with very restricted and limited ranges, and some known only as a single tree.

Many classic palaeoendemics, on the other hand, appear to be 'living fossils' which have remained essentially stable for millions of years. Two fine examples are the famous dawn redwood or water fir (*Metasequoia glyptostroboides*) (Fig. 8.2) and the maidenhair tree or ginkgo (*Ginkgo biloba*). The first of these was known as a fossil from the Tertiary (Chaney 1948, Zaleska 1950) *before* it was eventually discovered in the 1940s as a living tree. Today it is restricted in its native distribution to eastern Szechwan and north–eastern Hupeh in south–west China. The second, which also possesses a present–day natural or semi–natural distribution confined to one small region of China, is the sole surviving species of a family (the Ginkgoaceae) which was important in Jurassic times.

Complications

Unfortunately, this simple division into palaeoendemics and neoendemics, although important, will not in itself suffice. Many writers have thus found it necessary to develop much more complex classifications. At least three variations on the basic theme need to be recognised:

(1) There is evidence that under certain conditions a relict species, which could be a palaeoendemic or an old holoendemic that has remained unchanged until fairly recently, may experience a new burst of evolution and diversifi-

Table 8.1 The commonly occurring attribute states of the three main kinds of endemics (*after* Richardson 1978, Table 1). (Reproduced with permission from H. E. Street (ed.) 1978. *Essays in plant taxonomy.* © Academic Press Inc. (London) Ltd.)

Attribute	Endemic status		
	Neo-	Holo-	Palaeo-
taxonomically isolated	−	−	+
geographically isolated	−	+	+
polymorphic	−	+	−
derived characters	+	+	−
environment stable	±	−	+
ploidy level high	+/−	+/−	(+)/−
potential to expand area	+	+/−	−
age (recent +, old −)	+	+/−	(+)/−

cation. The resultant taxa may well resemble neoendemics in their basic attributes, even though they have developed from an old endemic. According to Wulff (1943), it was Rikli who first termed such taxa 'active epibiotics'. In these cases, variation probably arises through the process of adaptive radiation, which is the diversification of form in response to the pressures of different ecological habitats. This is a positive process in which genetical response to environmental

Figure 8.2   The dawn redwood or water fir (*Metasequoia glyptostroboides*) – a palaeoendemic taxon surviving in south-west China. It was first discovered as a living tree in 1941. (a) A living shoot picked from the tree in the month of June, i.e. while still expanding; (b) a fossil shoot, probably as shed in autumn, from Canada. (Plates kindly provided by Dr J. R. Flenley.)

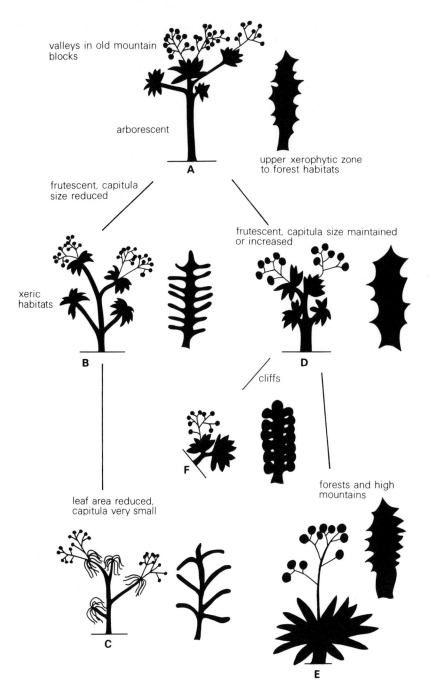

Figure 8.3 Adaptive radiation in growth-form, leaf shape and **capitulum** size in the Canarian members of the genus *Sonchus.* See text for full explanation (*after* Bramwell 1972, Fig. 7; © Linnean Society of London).

stimuli is the main factor. A particularly striking example is given by Bramwell (1972) in his study of the genus *Sonchus* in the Canary Islands (Fig. 8.3).

An initial growth form is represented by A in the diagram. This is based on *S. pinnata,* a tall (4 m) arborescent species found in the Canaries on several of the islands and also on Madeira. A similar species (*S. daltonii*) occurs in the Cape Verde group. Bramwell argues that 'this widespread distribution pattern along with the woody habit and large **capitula** may be considered as primitive characteristics' and that 'it is possible to derive the other major forms of *Sonchus* found in the Canaries from it'. The evolutionary trends shown in Figure 8.3 are closely correlated with different ecological conditions, each resulting in an adapted form fitted to a particular habitat. There are, therefore, distinctive species found in xeric habitats (B & C), in forested areas (D), up high mountains (E) and on cliffs (F).

(2) Endemism can also result through vicariance, a phenomenon already discussed in Chapter 7 with respect to disjunct distribution patterns. Vicarious species, or species-pairs, are those with the same parent type, but which have diverged through being geographically or ecologically isolated. In this instance, whether one treats the taxa in question as disjunct or endemic is immaterial and will depend on the point of view of the study in question.

Such vicarious evolution has been invoked by Bramwell, again working in the Canaries, to explain the distribution of endemic species in *Centaurea* sect. *Cheirolophus* subsect. *Flaviflorae* (Fig. 8.4). He suggests that species formation 'has probably taken place by fragmentation of a once widespread parent species' and he attributes this fragmentation to volcanic activity. Richardson (1978), however, questions this thesis, arguing that the different taxa have arisen by

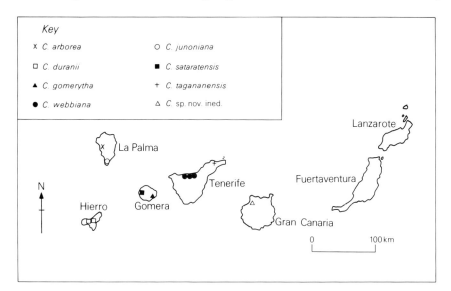

Figure 8.4   The distribution of species of *Centaurea* sect. *Cheirolophus* subsect. *Flaviflorae* in the Canary Islands (*after* Bramwell 1972, Fig. 10; © Linnean Society of London).

stepwise or repeated migration from extant stock. He thus agrees with Tryon (1971) that, although species of limited distribution in a source area tend to migrate to islands less frequently than those of broad distribution, when they actually do 'they are more likely to evolve into insular endemics'. In both theories, however, the rôle of geographical isolation is clearly evident.

(3) Thirdly, it is necessary to recognise the true holoendemic, with a range permanently restricted by physiological and ecological constraints. Good examples are provided by *Cupressus sargentii,* which is confined to the **serpentine soils** of California and Oregon, and *Fouquieria shrevei,* which grows on gypsum soils in northern Mexico (Daubenmire 1978). As Daubenmire comments, such endemics probably arise through a rare event, in which mutation or hybridisation produces from parent stock an individual pre-adapted to a very special environment that just happens to be in close proximity and open for colonisation. The geological age of such a special environment represents the maximum possible age for the endemics it supports, though they are probably much younger.

In addition to these three main variations on the basic division into neo-endemics and palaeoendemics, a number of authors have proposed certain specialised systems for the classification of endemics. For example, Favarger and Contandriopoulos (1961) have produced a classification based on cytology (see Chapter 9). They recognise four main groups:

(a) Palaeoendemics – isolated systematically, old, with little variation and not necessarily having arisen in their area of present survival;
(b) Schizoendemics – produced by 'gradual speciation', having a common origin and identical chromosome numbers;
(c) Patroendemics – narrow diploids which have given rise to widely distributed polyploids (see Ch. 9, pp. 126–7);
(d) Apoendemics – narrow polyploids which have arisen from widely distributed diploids.

However, in developing any classification of endemic types, it is vital to remember one simple fact, namely that, in many areas, the statistics on endemism and its significance in a flora tend to be exaggerated. The world is full of what we may unflatteringly call 'paper endemics', which are taxa that have been published and described as endemic to a certain region, but which, on detailed taxonomic revision, are found to be indistinguishable from more widely distributed taxa or which are simply forms of common taxa. In these cases, the endemic described is little more than a figment of the taxonomic imagination.

Van Balgooy (1971) has illustrated this danger fully in his detailed revision of certain genera for *Flora Malesiana.* For example, the genus *Canarium* was previously thought to contain forty-five species, all of which were endemic to this region. Now the genus is considered to be represented by nine species, only four of which are endemic. Such cases underline the fact that, in studying endemism, it is vital to have a very thorough attitude to taxonomic revision and to the improvement of distributional data. Perhaps the most basic principle in the study of endemism is the need for rigorous taxonomic treatments that are, however, sufficiently broad to allow for the revision of a genus on a world scale or, at least, on a large enough scale to embrace the main floristic region involved. The importance of the relationship between taxonomy and endemism has been fully explored by Richardson (1978).

The phenomenon of endemism

In general terms, the number of endemics in the northern hemisphere is lower than that in the southern hemisphere, although in both hemispheres their count is particularly reduced in the lands which were occupied during the Pleistocene by the cap of the continental glacier. The same is also true for mountains which were under ice, a fact suggesting that there is often a close relationship between the number of endemics and the geological age of the habitats that they occupy. Moreover, younger terrains tend to be characterised by neoendemics, whereas old islands and regions long free of glaciation possess palaeoendemic taxa also.

Further, it is well documented that the largest numbers of endemics are found on islands in the warmer regions of the world (which, however, possess more species anyway), and on isolated mountain groups. This fact underlines the intimate relationship that must exist between the phenomenon of endemism and high levels of geographical isolation. We shall conclude this chapter by examining the nature of this relationship both on islands and on isolated mountain ranges.

Islands. It is not without significance that the examples chosen to illustrate recent evolution within an older endemic flora came from an island setting, the Canaries. The study of endemism and island floras go hand in hand. Ever since Darwin's famous work on the Galápagos islands, islands and island groups have been regarded by many biologists as 'evolutionary laboratories', since they are more or less geographically isolated and have a wide range of unfilled niches just ripe for occupation through adaptive radiation, which can lead to a wealth of new endemic taxa. But, as pointed out by Wallace in his classic work *Island Life* published in 1880 (2nd edn 1892), it is important to distinguish two types of island. On the one hand, there are remote oceanic islands with what Greuter (1972) calls true 'island biosystems' and, on the other, there are those which were once connected with continental areas and which possess 'sub-continental biosystems'. An example of each type will illustrate the floristic differences between them.

The flora of Crete is of the sub-continental type. In mid-Tertiary times, it was an integral part of a continental area, its present-day mountains being remnants of an old mountain system that connected the Balkans with southern Anatolia.

According to Greuter (1972), on one estimate 132 species of vascular plants (around 9% of the total wild flora) are endemic, with some of them having arisen by the processes already outlined for the Canaries, namely adaptive radiation and vicariance. Others are genuine palaeoendemics which have now become extinct outside Crete, or patroendemics, in the sense of Favarger and Contandriopoulos (1961), being taxa which have remained unchanged in one place but have evolved further elsewhere, thus leaving the old forms as endemics. Such relict floras are highly characteristic of islands with a 'sub-continental biosystem'. On Crete, a large number of species, such as the Cretian pink (*Dianthus creticus*), appear to have undergone no evolutionary change since the upper Miocene. They form 'static populations' that are prevented from invading sizeable new areas and from colonising new habitats because they are highly specialised and unable to compete with other plants. Such floras, which were formerly part of a balanced continental flora, have evolved through a gradual impoverishment due to the extinction of species, counterbalanced to a certain extent by some evolution within the palaeoendemic flora and the establishment of occasional newcomers through dispersal.

Oceanic islands, in contrast, have always been independent of any large land mass, though they may lie fairly close to continents. Their floras, which have never formed an integral part of a continental flora, have been derived from outside and often from a diverse range of sources. The two main factors governing their composition are the age of the island in question and the degree of geographical isolation. The chief botanical features of such oceanic islands have been recognised since Darwin. For the most part, they possess fewer species in total but a higher percentage of endemics than equivalent continental areas, and the order of importance and representation of taxonomic groups are usually markedly different from the pattern on continents.

These characteristics are well exemplified by the flora of the Hawaiian Islands, which lie no less than 3900 miles (6275 km) from Japan and 2400 miles (3860 km) from America and which, in consequence, form the most isolated of all floristic regions. The total flora is small, with estimates ranging from the 216 genera and 1729 species and varieties of Fosberg (1948) to a more recent count of 238 genera by van Balgooy (1960). Moreover, it is a distinctive assemblage of genera *selected for their capacity to accomplish long-range dispersal*. According to Fosberg, their affinities are as follows: Indo–Pacific (40%), Austral (16·5%), American (18%), and Pantropic (12·5%), the remaining proportion being Boreal or unknown in origin. But the most telling fact is that on many estimates no less than 90% of native Hawaiian plants are endemic to the islands. Fosberg has actually concluded that this remarkable flora has probably evolved from as few as 272 original immigrants. If this were true, and granting that no part of the islands is deemed older than five million years, Good (1974) noted that this would mean, on average, the arrival and establishment of one species every 2000 years. According to Gillett (1972), in addition to vicariance (as in the palm genus *Pritchardia* and the genus *Viola*), natural hybridisation has been a significant evolutionary process in the later development of the flora.

In recent years, the whole subject of island biogeography has received much attention. In part, this has been generated by MacArthur and Wilson's *The theory of island biogeography* published in 1967. They argue that the number of species on an island is the result of a dynamic equilibrium between species arriving and colonising (which is a function of distance from source) and species becoming extinct (which is a function of island area). Their work has been reviewed by Stoddart (1977) and Gilbert (1980).

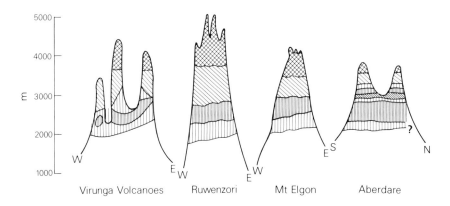

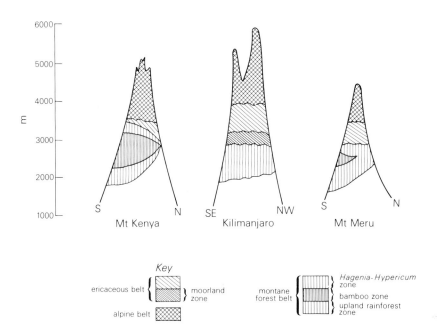

Figure 8.5 Schematic profiles showing the vegetation belts of certain East African mountains. The wettest side of each mountain is turned to the left. Only the vertical distances are drawn to scale (*after* Hedberg 1951, Fig. 4).

Mountains. Like oceanic islands, mountains – 'islands in the sky' – are noted for their endemic floras. Their isolation is not only geographical, but also ecological. The higher vegetation belts of one mountain are separated from the same belts on its nearest neighbour by the intervening lowland vegetation, which forms an ecological barrier between them (Fig. 8.5). In the higher vegetation belts, small populations are subject to a range of distinctive ecological pressures, which at the same time reduce the inflow of competitors from lower altitudes. Under these conditions, evolution is accelerated, with relicts differentiating through adaptive radiation and vicariance, and new endemics arising.

Such is the case in the remarkable Afro-alpine flora of the East African mountains, which has been intensively studied by Hedberg (1951, 1961, 1964, 1969). Of the 278 taxa of vascular plants recorded for this flora, no less than 81% are endemic to the East African region. These figures are remarkably close to those discussed for the flora of the Hawaiian Islands and clearly emphasise the similarities between the floras of isolated mountains and those of islands. At the generic level, the endemism drops to about 19% and the majority of Afro-alpine genera have their closest living relatives elsewhere. The most famous plants of these mountains are, perhaps, the 'tree lobelias' and the 'dendrosenecios', both of which exhibit an interesting series of adaptations to the distinctive climatic characteristics obtaining in the Afro-alpine belt, where every day is like summer and every night like winter.

One of the more important questions to consider in discussing endemism in mountain floras is the degree to which intermontane dispersal takes place, thus reducing the degree of isolation and providing an alternative explanation to

Table 8.2   The assumed dispersal capacity of seeds in 183 species of the Loma mountains and Tingi hills, West Africa (*after* Morton 1972, Table VI; © Linnean Society of London).

Mode of seed dispersal	Percentage of total flora
small seeds, possibly suited for dispersal by wind or on the feet and feathers of birds	22
succulent fruits; or seeds eaten by birds or bats	12
with effective means of wind dispersal (e.g. pappus)	6
	—
total adapted to some form of long-distance dispersal	40
	—
with sticky seeds or fruits, suited for dispersal by mammals*	9
inefficient means of wind dispersal suited for local transport only	6
explosive mechanisms	6
adapted for dispersal by ants	2
	—
total adapted for local dispersal	23
	—
with large seeds or fruits with no apparent means of dispersal	37
	—

*Perhaps also suitable for long-distance dispersal by birds.

evolution within relict floras which have become stranded on mountain tops through climatic change. In an analysis of the dispersal potential of the flora of one West African mountain group, the Loma mountains and the Tingi hills, Morton (1972) has shown that the flora has a distinctly limited capacity for such intermontane dispersal (see Table 8.2). Of the 183 montane species studied, less than half had any possible means of traversing appreciable distances and even among the rest, there were few equipped for a long journey. Morton argues that the present-day distribution of montane floras in Africa is the result of a complex series of factors, including climatic change which, during at least two cool periods (with a recession in temperature of between 4°C and 9°C), depressed the mountain vegetation belts by up to 1000 m, thus creating 'stepping stones' between the main mountain systems. This thesis is obviously linked with the idea of dispersal over short distances, and, moreover, he does allow for occasional long-distance dispersal by wind and birds. The rôle of climatic change in this context is hotly disputed, however, and the evidence for it is subject to problems of interpretation (Flenley 1979).

Further reading

Carlquist, S. 1974. *Island biology*. New York: Columbia University Press.
Favarger, C. and J. Contandriopoulos 1961. Essai sur l'endémism. *Bericht Schweiz. bot. Ges.* **71**, 384–406.
Gilbert, F. S. 1980. The equilibrium theory of island biogeography: fact or fiction? *J. Biogeog.* **7**, 209–35.
Hedberg, O. 1969. Evolution and speciation in a tropical high mountain flora. *Biol. J. Linn. Soc.* **1**, 135–48.
MacArthur, R. H. and E. O. Wilson 1967. *The theory of island biogeography*. Princeton, NJ: Princeton University Press.
Prentice, H. C. 1976. A study in endemism: *Silene diclinis*. *Biol. Conserv.* **10**, 15–30.
Richardson, I. B. K. 1978. Endemic taxa and the taxonomist. In *Essays in plant taxonomy*, H. E. Street (ed.), 245–62. London & New York: Academic Press.
Stebbins, G. L. and J. Major 1965. Endemism and speciation in the California flora. *Ecol. Monogr.* **35**, 1–35.
Valentine, D. H. (ed.) 1972. *Taxonomy, phytogeography and evolution*. London & New York: Academic Press. (Section **III**.)
Wallace, A. R. 1892. *Island life*, 2nd edn. London: Macmillan.

9 Genetics, plant geography and the plant kingdom

So far in this book, we have been concerned primarily to discuss the geographical distribution of conventional taxonomic units, especially the family, the genus and the species. We have linked, therefore, the disciplines of taxonomy and plant geography. Taxonomy is involved with the classification, description and naming of variation in organisms, whereas plant geography is the science which analyses the geographical expression of this variation in the plant kingdom. But our understanding of this variation is not confined to the traditional units, or models, of the taxonomist. In describing a species, a taxonomist is only setting up an hypothesis (Heywood 1973). There are, of course, hundreds or thousands of individuals with characteristics similar to those in the model. The hypothesis can only be tested by countless checking and rechecking in the field, herbarium and laboratory. We must now consider, therefore, the phytogeographical interest of variation below the level of conventional taxonomic units and the importance of the sequence:

individual $\longrightarrow$ population $\longrightarrow$ species $\longrightarrow$ genus $\longrightarrow$, etc.

The individual

The taxonomist works not only with the external appearance of an organism (the phenotype), but also with its genetic make-up (the genotype). This basic division was made clear by Johannsen in his selection experiments with the kidney bean (*Phaseolus vulgaris*) (Johannsen 1903, Grant 1975). In his experiments, two different types of variation were detected. The first stemmed from genotypic differences between separate lines of bean plants. The second was recognised because it became clear that any given genotype will give rise to a certain range of phenotypes, depending on environmental variables. The genotype is thus the sum total of genetic determinants and the phenotype is its particular character expression. We may further usefully distinguish the exo-phenotype, which is the set of externally expressed characters, such as leaf shape and flower colour, from the endo-phenotype, which comprises such internal features of the organism as chromosome form (Fig. 9.1) and biochemical products (Lewis & John 1963).

Traditional taxonomy has been above all concerned with discovering classifications of individuals based on the detailed study of exo-phenotypic variation. Likewise, plant geography has mainly concentrated on the geographical distribution of such external variation, expressed in the form of conventional taxonomic units. However, as we have already seen in Chapter 2, modern taxonomic methods are increasingly taking account of variation in both the endo-phenotype and the genotype.

For any individual organism, the general pattern of its development is laid down within certain channels by the sum total of the **genes** in the **zygote,** but the final phenotype will be the product of a very complex series of reactions between these genes and a wide range of both internal and external environmental factors (Grant 1975, Heywood 1967). Between primary gene action and the actual phenotypic expression, there are usually many developmental steps. In certain instances a single gene may be involved in the differentiation of several phenotypic characters (pleiotropy). More commonly, several genes may contribute to the production of a single character (polygeny). In both cases, however, the effects of gene action will be modified by environmental conditions. In many organisms the same genotype can produce a wide range of phenotypes through modification in different environments. Such variations are called phenotypic modifications and the phenomenon is known as phenotypic plasticity. It appears, however, that the capacity of phenotypes to respond to environmental influences is itself under genetic control (Bradshaw 1965).

In analysing the phytogeographical expression of such individual variation, it is clear that there must exist a very close relationship between plant geography and both genetics and ecology (Snaydon 1973).

The population

In nature, however, individuals tend to be grouped together in large numbers. Such groupings are termed populations. On the popular level this term refers to all the organisms, both plant and animal, which occupy a given geographical area or habitat. For the general taxonomist, on the other hand, a population is any group of individuals considered together at any one time because they have features in common. The population is thus defined by identifying blocks of individuals with similar phenotypic characters. But such populations are not necessarily gene pools or breeding populations in which individual members can exchange genes. Geneticists therefore recognise a more fundamental concept, that of the local breeding population. In this, the constituent individuals grow together in a defined geographical area and interbreed to form a common gene pool. The local breeding population is thus the unit of evolutionary change.

Such populations are obviously dynamic. They exhibit variation both between individuals within the population, and between populations. This variation is brought about by modifications due to the external environment, mutation and genetic recombination (Heywood 1967). The pattern of variation is largely determined by the breeding system.

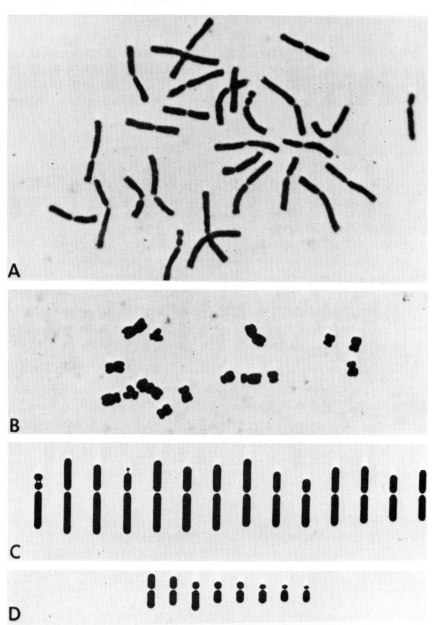

Figure 9.1 The somatic chromosome complements of two species of New Zealand angiosperms – (a) & (c) *Cockaynea gracilis* (Gramineae), 2*n*=28; (b) & (d) *Myosurus novae-zelandiae* (Ranunculaceae), 2*n*=16. Figures (a) & (b) are photomicrographs of mitotic metaphase. Figures (c) & (d) are idiograms, i.e. diagrammatic representations of the numerical and morphological characters of the basic **haploid** chromosome complement (*after* J. B. Hair *et al.* 1967; Figs 6, 12, 19 and 20. The figure presented here is in the form given by V. Grant 1975; Fig. 34 and is reproduced by courtesy of the New Zealand Department of Scientific and Industrial Research.)

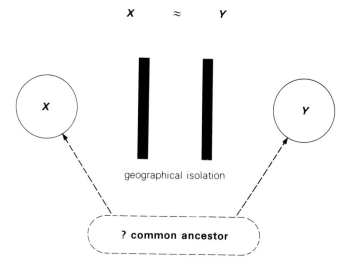

Figure 9.2 The case of two closely related but geographically isolated taxa, *X* and *Y*. See text for explanation (pp. 125–6).

As the subject of plant geography becomes more sophisticated, it is inevitable that greater attention will be paid to studies at the level of both the individual and the population. In this way, the analysis of geographical variation within the plant kingdom can be increasingly refined. Such a trend will be accompanied by detailed research into small-scale morphological and cytological variability, into anatomical, biochemical and genetic features, into pollination mechanisms and breeding systems and into the evidence of hybridity (Clapham 1972). Only by these means will it be possible to interpret satisfactorily the complex geographical expression of both genotypic and phenotypic variation.

To illustrate this point, we may refer to the simple theoretical case shown in Figure 9.2. The two taxa, *X* and *Y*, each occur in one site only and their sites are geographically isolated. Taxonomists are divided over the exact status of the taxa involved. Most regard *X* and *Y* as very closely related species that are probably vicariants which have evolved from a common ancestor and which differ only slightly in their external characters. Some have even described *Y* as a subspecies or variety of *X*. However, from the phytogeographical point of view, it is quite unimportant what taxonomists have called them or what status they are given. A more illuminating approach would be to consider the individuals at each site as comprising two geographically isolated populations with a significant overlap in their gene pools. A detailed research programme to test this hypothesis might then include:

(a) Biometric and small-scale morphological studies of the exo-phenotypic variation exhibited both within and between populations. Obviously, careful sampling procedures would have to be observed. Numerical taxonomy might then be employed.

(b) Detailed studies of the endo-phenotypic variation, including the karyotype. This is 'the phenotypic appearance of the somatic chromosomes' (Jackson 1971) and includes the numerical, morphological and anatomical characteristics of a somatic chromosome set (Fig. 9.1). The analysis of these characteristics comprises a branch of cytology (see Darlington 1963, and for methods in the study of chromosomes, Darlington & La Cour 1962, Dyer 1979).

(c) Carefully planned studies of genetic variation, including growth and hybridisation experiments using seed from individuals in both populations.

(d) Detailed studies of the ecological and biotic factors obtaining at the two sites involved.

Increasingly, therefore, the plant geographer's attention will be drawn to the individual, the population and the community as a whole. To him the species is an abstraction and the phytogeographical value of the species depends upon the amount of geographical information it conveys. As he wishes to refine this information, he will more and more strive to look at what lies beneath the traditional taxonomic categories and, in doing so, will begin to analyse variation at greater depths. None of these developments, however, will negate the value of the relationship that must persist between classical taxonomy and plant geography. At the world scale, and in respect of the higher taxonomic categories, this level of analysis will remain of enormous importance for a long time to come.

Polyploidy

The phytogeographical implications of one aspect of genetics, namely the phenomenon of polyploidy, have been recognised for a long time however (Cain 1944, Polunin 1960, Stebbins 1950, 1971). Polyploidy is the occurrence of multiple chromosome sets in an organism and it is widespread in the flowering plants, especially among perennial herbs.

The basic number of chromosomes (usually referred to as x) may occur as exact multiples in the different members of a group, thus creating what is termed a polyploid series ($2x$=diploid, $4x$=tetraploid, $5x$=pentaploid, $6x$= hexaploid and so on). A good example of such a polyploid series, with the basic number 6, is found in the alternate-leaved members of the golden saxifrage genus, *Chrysosplenium* (Saxifragaceae), in which *C. tetrandrum*, *C. wrightii* and *C. japonicum* of Japan have 24 somatic chromosomes, *C. alternifolium* of Europe has 48, *C. rosendahlii* of N. America has 96 and *C. iowense*, also of N. America, has about 120 (Hara 1972).

A large number of polyploids are formed as a result of the doubling of chromosomes of hybrids formed between separate species (**allopolyploidy**), or, at least, between the different races of single species. Often there is initially very little morphological difference between diploids and polyploids and they are included in a single species, or the polyploids are classified as subspecies and varieties. Only if the differences in morphology are striking and the polyploid is able to occupy a distinctive niche and establish itself in clearly differentiated populations will it be regarded as a separate species.

The phytogeographical significance of polyploids lies in the fact that they are often more vigorous than the corresponding diploid and may possess ecological advantages and increased competitive ability. These factors may allow them to replace their progenitors or to occupy new climatic or edaphic areas. In other words, polyploidy is an internal factor of distribution which may permit a species-complex to change and expand its geographical range. The different distributional histories of the diploid and tetraploid races of the horseshoe vetch *(Hippocrepis comosa)*, discussed at length in Chapter 6 (pp. 88–9), illustrate this point well and provide an excellent example of a polyploid which is able to colonise a far wider and more diverse range of habitats than the diploid. The general phytogeographical implications of polyploidy have been clearly outlined by Stebbins (1971, 179–90).

Genetic resource conservation

If it is right for this last chapter to emphasise the genetic and evolutionary basis of plant geography, it is even more appropriate to end the chapter, and the book, with a brief consideration of the crucial topic of genetic resource conservation, for this is a subject which concerns the future of the whole plant kingdom.

One of man's greatest resources is the genetic diversity of the plant kingdom (Table 9.1). In the past, he has drawn on this diversity to evolve new crop plants, drugs, chemical products and garden flowers, and it is the basic proposition of those concerned with genetic resource conservation that 'the diversity of the living kingdoms should be preserved, because that diversity is now, or is likely to be in the future, of direct value to man' (Heslop-Harrison 1974). The aim of genetic resource conservation is thus to preserve as wide a range of genotypes as possible so that we may avoid the dangers, especially evident in crop plants, of increasing evolutionary inflexibility, and protect the gene pools of plants and animals until their potential has been fully realised by man. What is more, unlike most of man's resources, genetic resources are self-renewing, so long as the plants and animals survive and are able to reproduce themselves.

Unfortunately, the plant kingdom is under threat and its survival is no longer guaranteed (Lucas & Synge 1978). Habitat destruction, industrialisation, the use of herbicides and pesticides, the pollution of air, land and water, the intensification and spread of agriculture, and man's population growth are inevitably and inexorably taking a heavy toll of the natural vegetation of the world. The speed of change is too rapid for plant communities to evolve and to adapt to the new environments created, many of which appear in any event to be intrinsically inimical to most forms of life. The Red Data Book for flowering plants, which is being compiled at the Royal Botanic Gardens, Kew, by Dr Ronald Melville, already lists some 20 000 species regarded as endangered. This is no less than about one twentieth of the whole group.

Against this background, the United Nations Conference on the Human Environment held in Stockholm in 1972 marked a significant landmark for

Table 9.1   A classification of the plant kingdom. (Reproduced from Harold C. Bold, *The plant kingdom,* 4th edn, © 1977, pp. 5, 7. Reprinted by permission of Prentice-Hall, Inc. Englewood Cliffs, New Jersey.)

Bold's classification (1977)		*Common names*	*Approximate number of living species*
Division 1	Chlorophycophyta	green algae	
Division 2	Euglenophycophyta	euglenoids	
Division 3	Charophyta	charophytes	
Division 4	Phaeophycophyta	brown algae	
Division 5	Rhodophycophyta	red algae	
Division 6	Chrysophycophyta	chrysophytes (diatoms, etc.)	20 000
Division 7	Pyrrhophycophyta	dinoflagellates fungi (*sensu lato*)	
Division 8	Cyanochloronta	blue-green algae	
Division 9	Schizonta	bacteria	
Division 10	Myxomycota	slime molds	
Division 11	Acrasiomycota	cellular slime molds fungi (*sensu stricto*)	
Division 12	Chytridiomycota	posteriorly uniflagellate fungi	50 000
Division 13	Oomycota	water molds and others	
Division 14	Zygomycota	bread molds and others	
Division 15	Ascomycota	sac fungi	
Division 16	Basidiomycota	club fungi	
Division 17	Deuteromycota	imperfect fungi	
Division 18	Hepatophyta	liverworts and hornworts*	6000
Division 19	Bryophyta	mosses	14 000
Division 20	Psilotophyta	psilotophytes	8
Division 21	Microphyllophyta	club mosses	1000
Division 22	Arthrophyta	horsetails and sphenopsids	10–25
Division 23	Pteridophyta	ferns	9500
Division 24	Cycadophyta	cycads	100
Division 25	Ginkgophyta	maidenhair tree (ginkgo)	1
Division 26	Coniferophyta	conifers	550
Division 27	Gnetophyta	—	71
Division 28	Anthophyta	flowering plants or angiosperms	300 000
		approximate total:	400 000

*Not to be confused with the hornwort genus (*Ceratophyllum*) of the flowering plants. 'Hornworts' here refers to the order Anthocerotales, a distinctive liverwort order sometimes raised to the rank of a class (Anthocerotopsida), thus equating it with the mosses and the liverworts.

genetic resource conservation. Happily, it was a topic which featured prominently in the recommendations for action (recommendations 39–45 of the *Declaration on the Human Environment*) which were agreed at the conference. The recommendations for plants advised the following (Heslop-Harrison 1973, 1974):

(a) the survey of plant genetic resources throughout the world, with appropriate programmes for exploration, research and collecting;
(b) the preparation of inventories of resources in the wild and in existing plant and seed collections, and the development of a comprehensive documentation system to handle information about what exists;
(c) the conservation of plant genetic resources in the field by the designation of reserves and the establishment of suitable management systems for them;
(d) the long-term conservation of plant genetic resources in living collections and in seed banks.

A plant geographer must be deeply concerned to support and help to carry out these recommendations. What we are witnessing is the progressive destruction of the space available for plants to develop their areas. Very soon many plant areas will have been lost for ever and the genetic resources of the plants with them. If we are unable to maintain the areas, let us at least protect the plants. And in the meantime, plant geography can continue to contribute by mapping, describing and interpreting the distribution of the plant resources of the world.

Further reading

Useful introductory texts are marked with an asterisk.

★Bold, H. C. 1977. *The plant kingdom*, 4th edn. Englewood Cliffs, NJ: Prentice-Hall.
Darlington, C. D. 1963. *Chromosome botany*, 2nd edn. London: George Allen & Unwin.
★Dyer, A. F. 1979. *Investigating chromosomes*. London: Edward Arnold.
Grant, V. 1975. *Genetics of flowering plants*. New York: Columbia University Press.
Heslop-Harrison, J. 1973. The plant kingdom: an exhaustible resource? (Sir William Wright Smith Memorial Lecture). *Trans. Bot. Soc. Edin.* **42**, 1–15.
★Heslop-Harrison, J. 1974. Genetic resource conservation: the end and the means. *J. R. Soc. Arts*, **1974**, 157–69.
★Heywood, V. H. 1976. *Plant taxonomy*, 2nd edn. London: Edward Arnold. (See particularly Chs 5 & 10).
Lucas, G. and H. Synge (eds) 1978. *The IUCN plant red data book*. Morges, Switz.: IUCN.
★Shorrocks, B. 1978. *The genesis of diversity*. London: Hodder & Stoughton.
Snaydon, R. W. 1973. Ecological factors, genetic variation and speciation in plants. In *Taxonomy and ecology*, V. H. Heywood (ed.), 1–29. London & New York: Academic Press.
Stebbins, G. L. 1971. *Chromosomal evolution in higher plants*. London: Edward Arnold. (See 179–90 for a full discussion of the phytogeographical significance of polyploidy.)

Glossary

These technical terms are printed in **bold** on their first appearance in the text.

allopatric
A term used to describe **taxa** which occupy mutually exclusive geographical ranges; in contrast to **sympatric,** the term for taxa with overlapping ranges.

allopolyploid
A term used to describe a **polyploid** *(q.v.)* derived by hybridisation between two different species with doubling of the **chromosome** number; in contrast to **autopolyploid,** the term for a **polyploid** derived from one **diploid** *(q.v.)* by multiplication of its **chromosome** sets.

autecology
The study of the inter-relationships between the *individual* organism and its environment, including other organisms; cf. **synecology.**

biocoenosis
A term introduced in 1877 by Karl Möbius in his famous paper on oyster beds to describe the internal relationships of living communities.

biogeocoenosis
A term introduced by Sukachev in 1947 to describe the sum total of ecological niches (both plant and animal) with their environment; cf. **ecosystem.**

capitulum
A headlike inflorescence of composite flowers, e.g. as in the daisy *(Bellis perennis)*; pl. **capitula.**

chromosomes
Small, deeply staining bodies, found in all cell nuclei, which largely determine the inheritable characters of organisms.

competition
The influences of one organism on another (**competitors**), resulting from the demand for resources (including geographical space) which are in short supply.

cosmopolitan
A term used to describe a **taxon** which is native to all continents (often excepting Antarctica) and which is widely distributed in each.

cultivar
A 'cultivated variety' which has arisen in a garden or a nursery and which possesses no wild population. Their nomenclature is ruled by 'The International Code of Nomenclature for Cultivated Plants'. No authority citation is required. The cultivar name is printed in normal type, not italics, and either follows immediately after the specific epithet in single quotes (e.g. the copper beech, *Fagus sylvatica* 'Purpurea') or has in front of it 'cv.' and no quotes (e.g. *F. sylvatica* cv. Purpurea).

cytology
The study of the structure and function of cells.

dioecious
A term meaning having the sexes on *different* plants.

diploid
An organism possessing two sets of **haploid chromosomes** in its nuclei.

ecosystem	A concept in which the living (biotic) community and the non-living (abiotic) environment are viewed as a functioning, integrated system. The main approach to study in modern **synecology.**
edaphic	Pertaining to soil, especially to those soil factors which influence organisms.
flora	A list of all the plant species living in a particular area or region at a particular time.
gene	A portion of a **chromosome** with the potential to influence some character or characters of an organism.
haploid	A term indicating the possession of a single **chromosome** set.
herbaceous	Of plant tissue, meaning soft rather than woody.
hexaploid	An organism possessing six sets of **haploid chromosomes** in its nuclei.
host	An organism on which a **parasite** is growing.
hybrid	An offspring the parents of which are genetically unalike (e.g. the service-tree of Fontainebleau, *Sorbus × latifolia,* a hybrid of two different species, namely *Sorbus aria* and *S. torminalis.* It was discovered in the Forêt de Fontainebleau some time before 1750).
isohyet	A cartographic line connecting points of equal rainfall.
isotherm	A cartographic line connecting points of equal temperature.
life-form	A type of growth form in plants usually indicating adaptive significance; the most famous system of life-form classification is that developed by Raunkiaer (1907, 1934).
Malesia	A modern term used by biogeographers to denote the islands on and between the Sunda and Sahul shelves (i.e. Maritime South East Asia). It was coined to avoid confusion with the State of Malaysia.
monophyletic	A term to describe a taxonomic group, usually above a species, the members of which have had the same immediate ancestors; cf. **polyphyletic.**
mutation	An abrupt change in an organism – morphological or bio-chemical – which is transmitted to offspring.
parasite	An organism which derives its food wholly or partially (hemiparasite) from other living organisms to which it is attached or within which it exists.
phenology	The study of the timing of periodic events such as leaf production, flowering, fruit formation and fruit dispersal.
phylogeny	The evolutionary history and relationships of groups of organisms.

pollen grain   A microspore (of a flowering plant or conifer) containing a mature or immature male gametophyte.

polyphyletic   A term used to describe the condition of a taxonomic group, usually a genus or higher, the members of which have come from different phyletic stocks, but through convergent or parallel evolution have been mistakenly considered a natural group; cf. **monophyletic.**

polyploid   An organism having a **chromosome** number which is a multiple, greater than two, of the **haploid** number of its group.

serpentine soils   Soils developed over serpentine rocks, usually very rich in magnesium (serpentine is a mineral, chemically a hydrous silicate of magnesium).

synecology   The study of living communities and their environmental relationships; cf. **autecology.**

taxon   A taxonomic group of any rank, such as a species, genus, family etc.; pl. **taxa.**

tetraploid   An organism possessing four sets of **haploid chromosomes** in its nuclei.

xeric   Dry, describing habitats or environmental conditions.

zygote   The cell produced by the union of two gametes; the fertilised egg.

Bibliography

Adams, C. C. 1902. Southeastern United States as a center of geographical distribution of fauna and flora. *Biol. Bull. Mar. Biol. Lab. Woods Hole* **3**, 115–31.

Andersen, S. Th. 1970. The relative pollen productivity and pollen representation of north European trees, and correction factors for tree pollen spectra. *Damn. Geol. Unders.*, ser. 11 **96**, 1–99.

Andersen, S. Th. 1973. The differential pollen productivity of trees and its significance for the interpretation of a pollen diagram from a forested region. In *Quaternary plant ecology*, H. J. B. Birks and R. G. West (eds), 109–15. Oxford: Blackwell Scientific.

Andrews, H. N. 1961. *Studies in paleobotany*. New York: John Wiley.

Balgooy, M. M. J. van 1960. Preliminary plant-geographical analysis of the Pacific. *Blumea* **10**, 385–430.

Balgooy, M. M. J. van 1966. Distribution maps of Pacific plants. In *Pacific plant areas*, C. G. G. J. van Steenis and M. M. J. van Balgooy (eds), 76–7, 96–7. 114–5, 122–3, 150–1, 170–1, 240–1, 248–9, 268–9. *Blumea*, Suppl. **5**, 1–312.

Balgooy, M. M. J. van 1971. Plant geography of the Pacific, as based on the distribution of Phanerogam genera. *Blumea*, Suppl. **6**. Leiden.

Balgooy, M. M. J. van 1976. Phytogeography. In *New Guinea vegetation*, K. Paijmans (ed.), 1–22. Canberra: ANU Press/CSIRO.

Banks, H. P. 1970. *Evolution and plants of the past*. Belmont, Calif.: Wadsworth.

Bell, E. A. and C. S. Evans 1978. Biochemical evidence of a former link between Australia and the Mascarene Islands. *Nature* **273**, 295–6.

Berry, E. W. 1924. Age and area as viewed by the paleontologist. *Am. J. Bot.* **11**, 547–57.

Birks, H. J. B. 1973. Modern macrofossil assemblages in lake sediments in Minnesota. In *Quaternary plant ecology*, H. J. B. Birks and R. G. West (eds), 173–89. Oxford: Blackwell Scientific.

Bold, H. C. 1977. *The plant kingdom*, 4th edn. Englewood Cliffs, NJ: Prentice-Hall.

Bradshaw, A. D. 1965. Evolutionary significance of phenotypic plasticity in plants. *Adv. Genet.* **13**, 115–55.

Bramwell, D. 1972. Endemism in the flora of the Canary Islands. In *Taxonomy, phytogeography and evolution*, D. H. Valentine (ed.), 141–59. London & New York: Academic Press.

Brown, C. A. 1960. *Palynological techniques*. Baton Rouge: Louisiana State University Press.

Butzer, K. W. 1964. *Environment and archaeology*. London: Methuen.

Cadbury, D. A., J. G. Hawkes and R. C. Readett 1971. *A computer-mapped flora. A study of the county of Warwickshire*. London & New York: Academic Press.

Cain, S. A. 1944. *Foundations of plant geography*. New York: Harper & Row. (Reprinted 1974. New York: Hafner).

Candolle, A. de 1855. *Géographie botanique raisonnée ou exposition des faits principaux et des lois concernant la distribution géographique des plantes de l'époque actuelle*. Vols I & II. Paris: Masson and Geneva: J. Kessmann.

Carlquist, S. 1974. *Island biology*. New York: Columbia University Press.

Challice, J. S. and M. N. Westwood 1973. Numerical taxonomic studies of the genus *Pyrus* using both chemical and botanical characters. *Bot. J. Linn. Soc.* **67,** 121–48.

Chaney, R. W. 1940a. Bearing of forests on the theory of continental drift. *Scient. Mon.* Dec. 1940, 489–99.

Chaney, R. W. 1940b. Tertiary forests and continental history. *Bull. Geol. Soc. Am.* **51,** 469–86.

Chaney, R. W. 1948. The bearing of the living *Metasequoia* on problems of Tertiary paleobotany. *Proc. Nat. Acad. Sci.* **34.** Washington.

Clapham, A. R. 1972. Questions answered and unanswered. In *Taxonomy, phytogeography and evolution,* D. H. Valentine (ed.), 397–9. London & New York: Academic Press.

Clapham, A. R., T. G. Tutin and E. F. Warburg 1962. *Flora of the British Isles,* 2nd edn. Cambridge: Cambridge University Press.

Cormack, R. M. 1971. A review of classification. *J. R. Statist. Soc. A* **134,** 321–67.

Couper, R. A. 1960. Southern Hemisphere Mesozoic and Tertiary Podocarpaceae and Fagaceae, and their palaeographic significance. *Proc. R. Soc., Ser B* **152,** 491–500.

Cox, A. (ed.) 1973. *Plate tectonics and geomagnetic reversals.* San Francisco: W. H. Freeman.

Cox, C. B., I. N. Healey and P. D. Moore 1976. *Biogeography: an ecological and evolutionary approach,* 2nd edn. Oxford: Blackwell.

Croizat, L. 1952. *Manual of phytogeography.* The Hague: Junk.

Croizat, L. 1958. *Panbiogeography.* Caracas: the author.

Croizat, L. 1960. *Principia botanica.* Caracas: the author.

Dagnélie, P. A. 1960. Contribution à l'étude des communautés végétales par l'analyse factorielle. *Bull. Serv. Carte Phytogéogr. (Sér. B)* **5,** 7–71; 93–195.

Dandy, J. E. 1969. *Watsonian vice-counties of Great Britain.* London: Ray Society.

Darlington, C. D. 1963. *Chromosome botany,* 2nd edn. London: George Allen & Unwin.

Darlington, C. D. and L. F. La Cour 1976. *The handling of chromosomes,* 6th edn. London: George Allen & Unwin.

Darwin, C. 1859. *On the origin of species.* London: John Murray. (The 6th edn of 1872 is the edition most frequently cited.)

Daubenmire, R. F. 1974. *Plants and environment. A textbook of autecology,* 3rd edn. New York: John Wiley.

Daubenmire, R. F. 1975. Floristic plant geography of eastern Washington and northern Idaho. *J. Biogeog.* **2,** 1–18.

Daubenmire, R. F. 1978. *Plant geography, with special reference to North America.* New York & London: Academic Press.

Davis, M. B. 1967. Pollen deposition in lakes as measured by sediment traps. *Bull. Geol. Soc. Am.* **78,** 849–58.

Davis, P. H. and V. H. Heywood 1963. *Principles of angiosperm taxonomy.* Edinburgh & London: Oliver & Boyd.

Dietz, R. S. and J. C. Holden 1972. The breakup of Pangaea. In *Continents adrift* (readings from *Scientific American,* J. Tuzo Wilson, ed.), 102–13. San Francisco: W. H. Freeman. (See also: *Scient. Am.* 1968, **223,** 30–41.)

Drude, O. 1890. *Handbuch der Pflanzengeographie.* Stuttgart: J. Engelhorn. (French edn 1897. *Manuel de géographie botanique.* Paris.)

Du Toit, A. L. 1937. *Our wandering continents.* Edinburgh & London: Oliver & Boyd.

Dyer, A. F. 1979. *Investigating chromosomes.* London: Edward Arnold.

Edees, E. S. 1972. *Flora of Staffordshire: flowering plants and ferns.* Newton Abbot: David & Charles.

Embleton, C. and C. A. M. King 1968. *Glacial and periglacial geomorphology.* London: Edward Arnold.

Engler, A. 1879–82. *Versuch einer Entwicklungsgeschichte der Pflanzenwelt, insbesondere der Florengebiete seit der Tertiärperiode,* Vols I & II. Leipzig: Engelmann.

Engler, A. and L. Diels 1936. *Syllabus der Pflanzenfamilien.* Auf. 11. Berlin: Verlag von Gebrüder Borntraeger.

Erdtman, G. 1943. *An introduction to pollen analysis.* Waltham, Mass.: Chronica Botanica.

Evans, P. 1971. Towards a Pleistocene time-scale. In *The Phanerozoic time-scale – a supplement* (Spec. Publ. Geol. Soc. Lond. **5**), 123–356. London: The Geological Society.

Faegri, K. and J. Iversen 1974. *Textbook of pollen analysis,* 3rd edn by K. Faegri. Oxford: Blackwell Scientific.

Favarger, C. and J. Contandriopoulos 1961. Essai sur l'endémism. *Ber. Schweiz. Bot. Ges.* **71**, 384–406.

Fearn, G. M. 1972. The distribution of intraspecific chromosome races of *Hippocrepis comosa* L. and their phytogeographical significance. *New Phytol.* **71**, 1221–5.

Fitch, W. 1851. *Victoria regia or, illustrations of the Royal water-lily, in a series of figures chiefly made from specimens flowering at Syon and at Kew.* London: Reeve & Benham.

Flenley, J. R. 1973. The use of modern pollen rain samples in the study of the vegetational history of tropical regions. In *Quaternary plant ecology,* H. J. B. Birks and R. G. West (eds), 131–41. Oxford: Blackwell Scientific.

Flenley, J. R. 1979. *The equatorial rain forest: a geological history.* London: Butterworths.

Flint, R. F. 1971. *Glacial geology and Quaternary geology.* New York: John Wiley.

Florin, R. 1963. The distribution of conifer and taxad genera in time and space. *Acta Hort. Berg.* **20**, 121–312.

Fosberg, F. R. 1948. Derivation of the flora of the Hawaiian Islands. In *Insects of Hawaii.* Vol. I., E. C. Zimmerman (ed.), 107–19. Honolulu: University of Hawaii Press.

Fosberg, F. R. 1976. Geography, ecology and biogeography. *Ann. Assoc. Am. Geogrs* **66**, 117–28.

Fournier, P. 1946. *Les quatre flores de la France.* Paris: P. Lechevalier.

Frenzel, B. 1968. The Pleistocene vegetation of northern Eurasia. *Science* **161**, 637–49.

Fryxell, P. A. 1967. The interpretation of disjunct distributions. *Taxon* **16**, 316–24.

Gilbert, F. S. 1980. The equilibrium theory of island biogeography: fact or fiction? *J. Biogeog.* **7**, 209–35.

Gillett, G. W. 1972. The rôle of hybridization in the evolution of the Hawaiian Flora. In *Taxonomy, phytogeography and evolution,* D. H. Valentine (ed.), 205–19. London & New York: Academic Press.

Gilmour, J. S. L. (ed.) 1972. *Thomas Johnson. Botanical journeys in Kent and Hampstead.* Pittsburgh: The Hunt Botanical Library.

Gleason, H. A. 1924. Age and area from the viewpoint of phytogeography. *Am. J. Bot.* **11**, 541–6.

Gleason, H. A. and A. Cronquist 1964. *The natural geography of plants.* New York: Columbia University Press.

Godwin, H. 1975. *The history of the British Flora. A factual basis for phytogeography,* 2nd edn. Cambridge: Cambridge University Press.

Godwin, H. and P. A. Tallantire 1951. Studies in the post-glacial history of British vegetation. XII. Hockham Mere, Norfolk. *J. Ecol.* **39,** 285–301.

Good, R. 1936. On the distribution of the lizard orchid (*Himantoglossum hircinum* Koch). *New Phytol.* **35,** 142–70.

Good, R. 1974. *The geography of the flowering plants,* 4th edn. London: Longman.

Goodman, L. A. and W. H. Kruskal 1954. Measures of association for cross-classifications. *J. Am. Statist. Assoc.* **49,** 732–64.

Grant, V. 1975. *Genetics of flowering plants.* New York: Columbia University Press.

Gray, A. 1846. Analogy between the flora of Japan and that of the United States. *Am. J. Sci. Arts* **52,** 135–6.

Gray, A. 1859. Diagnostic characters of new species of phanerogamous plants, collected in Japan by Charles Wright, Botanist of the U.S. North Pacific Exploring Expedition. (Published by request of Captain John Rodgers, Commander of the Expedition.) With observations upon the relations of the Japanese flora to that of North America, and of other parts of the Northern Temperate Zone. *Mem. Am. Acad. Arts Sci.* II, **6,** 377–452.

Greenman, J. M. 1925. The age and area hypothesis with special reference to the flora of tropical America. *Am. J. Bot.* **12,** 189–93.

Greuter, W. 1972. The relict element of the flora of Crete and its evolutionary significance. In *Taxonomy, phytogeography and evolution,* D. H. Valentine (ed.), 161–77. London & New York: Academic Press.

Gribbin, J. (ed.) 1978. *Climatic change.* Cambridge: Cambridge University Press.

Grieg-Smith, P. 1978. *Quantitative plant ecology,* 3rd edn. Oxford: Blackwell Scientific.

Grisebach, A. 1872. *Die Vegetation der Erde nach ihrer Klimatischen Anordnung.* Leipzig: Engelmann.

Grisebach, A. 1880. *Gesammelte Abhandlungen und kleinere Schriften zur Pflanzengeographie.* Leipzig: Engelmann.

Gunn, C. R. and J. V. Dennis 1976. *World guide to tropical drift seeds and fruits.* New York: Demeter Press.

Guppy, H. B. 1906. *Observations of a naturalist in the Pacific between 1896 and 1899.* London: Macmillan.

Hagmeier, E. M. and C. D. Stults 1964. A numerical analysis of the distribution patterns of North American mammals. *System. Zool.* **13,** 125–55.

Hair, J. B., E. J. Beuzenberg and B. Pearson 1967. Contributions to a chromosome atlas of the New Zealand flora. IX. Miscellaneous Families. *N.Z.J. Bot.* **5,** 185–96.

Hall, J. B., A. J. Morton and S. S. Hooper 1976. Application of principal components analyses with constant character number in a study of the *Bulbostylis/Fimbristylis* (Cyperaceae) complex in Nigeria. *Bot. J. Linn. Soc.* **73,** 303–15.

Hallam, A. 1972. Continental drift and the fossil record. *Scient. Am.* **227,** 56–70.

Hallam, A. 1973. *A revolution in the earth sciences.* Oxford: Oxford University Press.

Hanks, S. L. and D. E. Fairbrothers 1976. Palynotaxonomic investigation of *Fagus* L. and *Nothofagus* Bl.: light microscopy, scanning electron microscopy, and computer analyses. In *Botanical systematics* **1,** V. H. Heywood (ed.), 1–141. London & New York: Academic Press.

Hara, H. 1972. Corresponding taxa in North America, Japan and the Himalayas. In *Taxonomy, phytogeography and evolution,* D. H. Valentine (ed.), 61–72. London & New York: Academic Press.

Harrison, C. M. 1971. Recent approaches to the description and analysis of vegetation. *Trans. Inst. Br. Geogrs* **52,** 113–27.

Hartigan, J. A. 1967. Representation of similarity matrices by trees. *J. Am. Statist. Assoc.* **62,** 1140–58.

Havinga, A. J. 1964. Investigation into the differential corrosion susceptibility of pollen and spores. *Pollen spores* **4,** 621–35.

Heather, D. C. 1979. *Plate tectonics.* London: Edward Arnold.

Hedberg, O. 1951. Vegetation belts of the East African mountains. *Svensk. Bot. Tid.* **45,** 140–202.

Hedberg, O. 1961. The phytogeographical position of the afroalpine flora. In *Recent advances in botany,* 914–9, Toronto: Toronto University Press.

Hedberg, O. 1964. Features of afroalpine ecology. *Acta Phytogeogr. Suec.* **49,** 1–144.

Hedberg, O. 1969. Evolution and speciation in a tropical high mountain flora. *Biol. J. Linn. Soc.* **1,** 135–48.

Heslop-Harrison, J. 1973. The plant kingdom: an exhaustible resource? (Sir William Wright Smith Memorial Lecture). *Trans Bot. Soc. Edinb.* **42,** 1–15.

Heslop-Harrison, J. 1974. Genetic resource conservation: the end and the means. *J. R. Soc. Arts* 1974: 157–69.

Hess, H. H. 1962. History of ocean basins. In *Petrological studies: a volume in honor of A. F. Buddington,* A. E. J. Engel, H. L. James and B. F. Leonard (eds), 599–620. New York: Geol. Soc. Am.

Heywood, V. H. (ed.) 1968. *Modern methods in plant taxonomy* (Bot. Soc. Br. Isles Conf. Report **10**). London & New York: Academic Press.

Heywood, V. H. (ed.) 1973. *Taxonomy and ecology.* London & New York: Academic Press.

Heywood, V. H. 1974. Systematics – the stone of Sisyphus. *Biol. J. Linn. Soc.* **6,** 169–78.

Heywood, V. H. 1976. *Plant taxonomy* 2nd edn. London: Edward Arnold.

Heywood, V. H. (ed.) 1978. *Flowering plants of the world.* Oxford: Oxford University Press.

Holland, P. G. 1978. An evolutionary biogeography of the genus *Aloë. J. Biogeog.* **5,** 213–26.

Holmes, A. 1931. Radioactivity and earth movements. *Trans Geol. Soc. Glasgow* **18,** 559–606.

Holmquist, C. 1962. The relict concept. *Oikos* **13,** 262–92.

Hooker, J. D. 1853. *The botany of the Antarctic Voyage of H.M. Discovery Ships* Erebus *and* Terror *in the years 1839–1843. Flora Novae-Zelandiae,* Vol. II, part I, *Flowering Plants.* London: Reeve.

Hultén, E. 1937. *Outline of the history of Arctic and Boreal biota during the Quaternary Period.* Stockholm: Bokfoerlags Aktaibolaget Thule.

Hultén, E. 1968. *Flora of Alaska and neighbouring territories.* Stanford, Calif.: Stanford University Press.

Humboldt, F. H. A. von and A. J. A. Bonpland 1805. *Essai sur la géographie des plantes: accompagné d'un tableau physique des régions équinoxiales.* Paris: Levrault, Schoell & Compagnie.

Hutchinson, J. and J. M. Dalziel 1954–68. *Flora of West Tropical Africa,* 2nd edn, R. W. J. Keay and F. N. Hepper (eds). London: Crown Agents.

Illies, J. 1974. *Introduction to zoogeography.* London: Macmillan.

Iversen, J. 1944. *Viscum, Hedera* and *Ilex* as climatic indicators. *Geol. Fören. Stockh. Förh.* **66,** 463.

Iversen, J. 1973. The development of Denmark's nature since the last glacial. *Danm. Geol. Unders.* Raekke **5** (7C).

Jackson, R. C. 1971. The karyotype in systematics. *Ann. Rev. Ecol. System.* **2,** 327–68.

Jardine, N. 1971. Patterns of differentiation between human local populations. *Phil. Trans. R. Soc. Lond. B* **263,** 1–33.

Jardine, N. 1972. Computational methods in the study of plant distributions. In *Taxonomy, phytogeography and evolution,* D. H. Valentine (ed.), 381–93. London & New York: Academic Press.

Jardine, N. and R. Sibson 1971. *Mathematical taxonomy.* New York: John Wiley.

Jeffrey, C. 1968. *An introduction to plant taxonomy.* London: J. & A. Churchill. (2nd edn 1977: Edward Arnold.)

Jeffrey, C. 1977. *Biological nomenclature,* 2nd edn. London: Edward Arnold.

Johannsen, W. 1903. *Ueber Erblichkeit in Populationen und in reinen Linien.* Jena: Gustav Fischer.

Johnson, T. 1629. *Iter Plantarum investigationis ergo susceptum a decem sociis, in Agrum Cantianum.* London: unidentifiable printer's device only (see Gilmour 1972).

Johnson, L. A. S. and B. G. Briggs 1975. On the Proteaceae – the evolution and classification of southern family. *Bot. J. Linn. Soc.* **70,** 83–182.

Jones, S. B. and A. E. Luchsinger 1979. *Plant systematics.* New York: McGraw-Hill.

Juniper, B. E., A. J. Gilchrist, G. C. Cox and P. R. Williams 1970. *Techniques for plant electron microscopy.* Oxford: Blackwell Scientific.

Kalkman, C. and P. Smit (eds) 1979. Rijksherbarium: 1829–1979. *Blumea* **25,** 1–140.

Kanis, A. 1968. A revision of the Ochnaceae of the Indo-Pacific area. *Blumea* **16,** 1–82.

Kausik, S. B. 1943. The distribution of the Proteaceae: past and present. *J. Ind. Bot. Soc.* **22,** 105–23.

Keates, J. S. 1973. *Cartographic design and production.* London: Longman.

Keast, D. 1971. Continental drift and the biota of the southern continents. *Q. Rev. Biol.* **46,** 335–78.

Kingdon Ward, F. 1937. *Plant hunter's paradise.* London: Jonathan Cape.

Kormondy, E. J. 1976. *Concepts of ecology,* 2nd edn. Englewood Cliffs, NJ: Prentice-Hall.

Kummel, B. and D. Raup (eds) 1965. *Handbook of paleontological techniques.* San Francisco: W. H. Freeman.

Larsen, S. S. 1975. Pollen morphology of Thai species of *Bauhinia* (Caesalpiniaceae). *Grana* **14,** 114–31.

Lauer, W. and H.-J. Klink 1978. *Pflanzengeographie.* Darmstadt: Wissenschaftliche Buchgesellschaft.

Lebrum, J.-P. and A. L. Stork 1977. *Index des cartes de répartition. Plantes vasculaires d'Afrique (1935–76).* Geneva: Conservatoire et Jardin botanique.

Lemée, G. 1967. *Précis de biogéographie.* Paris: Masson.

Lemmon, K. 1968. *The golden age of plant hunters.* London: Phoenix House.

Le Pichon, X., J. Francheteau and J. Bonnin 1973. *Plate tectonics.* Amsterdam: Elsevier.

Levitt, J. 1972. *Responses of plants to environmental stresses.* New York & London: Academic Press.

Lewis, K. R. and B. John 1963. *Chromosome marker.* Boston: Little, Brown & Co.

Li, H.-L. 1952. Floristic relationships between eastern Asia and eastern North America. *Trans Am. Phil. Soc.* **42,** 371–429.

Lind, E. M. and M. E. S. Morrison 1974. *East African vegetation.* London: Longman.

Lindley, J. 1837. *Victoria regia.* London: Shakespeare Press.

Lindley, J. 1838. *Victoria regia. Bot. Reg.* **24,** Misc. Not. 13.

Livingstone, D. A. 1975. Late Quaternary climatic change in Africa. *Ann. Rev. Ecol. System.* **6**, 332–41.

Livingstone, J. 1819. Chinese botany. *The Indo-Chinese Gleaner* **2**, 130.

Lucas, G. and H. Synge (eds) 1978. *The IUCN plant red data book.* Morges, Switz.: IUCN.

Lyell, C. 1853. *Principles of geology,* 9th edn. London: John Murray.

MacArthur, R. H. 1972. *Geographical ecology.* New York: Harper & Row.

MacArthur, R. H. and E. O. Wilson 1967. *The theory of island biogeography.* Princeton, NJ: Princeton University Press.

Masclef, A. 1888. Contributions nouvelles à la flore des collines d'Artois (Cambrésis, Artois, Haut-Boulonnais). *J. Bot., Paris* **2**, 271–5, 304–8, 324–8, 342–8, 359–68.

Matthews, J. R. 1955. *Origin and distribution of the British flora.* London: Hutchinson.

Mayr, E. 1963. *Animal species and evolution.* Cambridge, Mass.: Harvard University Press.

McQueen, D. R. 1969. Macroscopic plant remains in recent lake sediments. *Tuatara* **17**, 13–9.

Meusel, H. 1969. Beziehungen in der Florendifferenzierung von Eurasien und Nordamerika. *Flora,* Abt. B, **158**, 537–64.

Meusel, H., E. Jager and E. Winert 1965. *Vergleichende Chorologie der Zentral-europaischen Flora. Karten.* Jena: Gustav Fischer.

Meusel, H. and R. Schubert 1971. Beitrage zur Pflanzengeographie des Westhimaljas. I. Teil: Die Arealtypen. *Flora* **160**, 137–94.

Möbius, K. 1877. *Die Auster und die Austernwirtschaft.* Berlin: Wiegundt, Hempel & Parey. (Trans. *Rep. U.S. Comm. Fish.* 1880, 683–751.)

Moore, P. D. and J. A. Webb 1978. *An illustrated guide to pollen analysis.* London: Hodder & Stoughton.

Morton, J. K. 1972. Phytogeography of the West African mountains. In *Taxonomy, phytogeography and evolution,* D. H. Valentine (ed.). 221–36. London & New York: Academic Press.

Nelson, E. C. 1978. Tropical drift fruits and seeds on coasts in the British Isles and western Europe. I. Irish beaches. *Watsonia* **12**, 103–12.

Nicholson, R. J. 1951. A note on hollow curves. *New Phytol.* **50**, 138–9.

Nilsson, O. and Gustafsson, L.-Å. 1976. Projekt Linné rapporterar 14–28. *Svensk Bot. Tidskr.* **70**, 211–24.

Ødum, S. 1968. Udbredelsen af traeer og buske i Danmark. *Bot. Tidsskr.* **64**, 1–118.

Ozenda, P. 1964. *Biogéographie végétale.* Paris: Doin.

Packham, J. R., P. H. Oswald, F. H. Perring, C. A. Sinker and I. C. Trueman 1979. Preparing a new Flora of the Shropshire region using a federal system of recording. *Watsonia* **12**, 239–47.

Pears, N. 1977. *Basic biogeography.* London: Longman.

Pearsall, W. H. 1950. *Mountains and moorlands.* London: Collins.

Perring, F. H. 1968. *Critical supplement to the atlas of the British flora.* London: Nelson.

Perring, F. H. and S. M. Walters 1962. *Atlas of the British flora.* London: Nelson.

Phinney, R. A. (ed.) 1968. *History of the Earth's crust.* Princeton, NJ: Princeton University Press.

Pigott, C. D. 1956. The vegetation of Upper Teesdale in the North Pennines. *J. Ecol.* **44,** 545–86.

Pigott, C. D. and S. M. Walters 1954. On the interpretation of the discontinuous distributions shown by certain British species of open habitats. *J. Ecol* **42,** 95–116.

Polunin, N. 1960. *Introduction to plant geography and some related sciences.* London: Longman.

Praeger, R. L. 1896. On the botanical subdivision of Ireland. *Ir. Nat.* **5,** 29–38.

Prentice, H. C. 1976. A study in endemism: *Silene diclinis. Biol. Conserv.* **10,** 15–30.

Proctor, M. C. F. 1967. The distribution of British liverworts: a statistical analysis. *J. Ecol.* **55,** 119–35.

Radford, A. E., W. C. Dickison, J. R. Massey and C. R. Bell 1974. *Vascular plant systematics.* New York: Harper & Row.

Raunkiaer, C. 1907. *Planterigets Livsformer og deres Betydning for Geografien.* Copenhagen.

Raunkiaer, C. (ed. A. G. Tansley) 1934. *The life forms of plants and statistical plant geography.* Oxford: Clarendon Press.

Raup, H. M. 1941. Botanical problems in Boreal America. *Bot. Rev.* **7,** 147–248.

Raven, C. E. 1942. *John Ray, naturalist.* Cambridge: Cambridge University Press.

Raven, P. H. and D. I. Axelrod 1972. Plate tectonics and Australasian paleobio-geography. *Science* **176,** 1379–86.

Raven, P. H. and D. I. Axelrod 1974. Angiosperm biogeography and past continental movements. *Ann. Mo. Bot. Gdn.* **61,** 539–673.

Ray, J. 1660. *Catalogus Plantarum circa Cantabrigiam nascentium.* Cambridge: printed by John Field for William Nealand.

Richardson, I. B. K. 1978. Endemic taxa and the taxonomist. In *Essays in plant taxonomy* H. E. Street (ed.), 245–62. London & New York: Academic Press.

Ridley, H. N. 1930. *The dispersal of plants throughout the world.* Ashford: L. Reeve

Robinson, A., R. Sale and J. Morrison 1978. *Elements of cartography,* 4th edn. New York: John Wiley.

Roblin, H. S. 1969. *Map projections.* London: Edward Arnold.

Rompaey, E. van and L. Delvosalle 1972. *Atlas de la Flore Belge et Luxembourgeoise.* Brussels: Jardin botanique national de Belgique.

Rose, F, and P. W. James 1974. Regional studies on the British lichen flora 1. The corticolous and lignicolous species of the New Forest, Hampshire. *Lichenologist* **6,** 1–72.

Ross, C. A. (ed.) 1976. *Paleobiogeography.* Stroudsburg, Pennsylvania: Dowden, Hutchinson & Ross.

Ross Mackay, J. 1949. Dotting the dot map. *Surveying & Mapping* **9,** 3–10.

Rothmäler, W. 1950. *Allgemeine Taxonomie und Chorologie der Pflanzen.* Jena: W. Gronau.

Runcorn, S. K. (ed.) 1962. *Continental drift.* London & New York: Academic Press.

Ryvarden, L. 1971. Studies in seed dispersal. I. Trapping of diaspores in the alpine zone of Finse, Norway. *Norw. J. Bot.* **18,** 215–26.

Schimper, A. F. W. 1903. *Plant geography upon a physiological basis* (trans. W. R. Fisher). Oxford: Clarendon Press. (Reprinted 1960. New York: Harper & Row).

Schomburgk, R. 1838. Dr Robert H. Schomburgk's description of *Victoria Regina,* Gray. Plate XV. *Mag. Zool. Bot.* **2,** 440–2.

Schonlands, S. 1924. On the theory of age and area. *Ann. Bot.* **38,** 453–72.

Schouw, J. F. 1823. *Grundzüge einer allgemeinen Pflanzengeographie.* Berlin: G. Reimer.

Seddon, B. 1971. *Introduction to biogeography*. London: Duckworth.

Selander, S. 1950. Floristic phytogeography of south-western Lule Lappmark. *Acta Phytogeogr. Suec.* **27,** 1–200.

Sheail, J. and T. C. E. Wells 1980. The Marchioness of Huntley: the written record and the herbarium. *Biol. J. Linn. Soc.* **13,** 315–30.

Shimwell, D. W. 1971. *The description and classification of vegetation*. London: Sidgwick & Jackson.

Shorrocks, B. 1978. *The genesis of diversity*. London: Hodder & Stoughton.

Sinnott, E. W. 1924. Age and area and the history of species. *Am. J. Bot.* **11,** 573–8.

Smith, P. M. 1976. *The chemotaxonomy of plants*. London: Edward Arnold.

Snaydon, R. W. 1973. Ecological factors, genetic variation and speciation in plants. In *Taxonomy and ecology*, V. H. Heywood (ed.), 1–29. London & New York: Academic Press.

Sneath, P. H. A. and R. R. Sokal 1973. *Numerical taxonomy*. San Francisco: W. H. Freeman.

Sokal, R. R. and N. L. Oden. 1978a. Spatial autocorrelation in biology. 1. Methodology. *Biol. J. Linn. Soc.* **10,** 199–228.

Sokal, R. R. and N. L. Oden 1978b. Spatial autocorrelation in biology. 2. Some biological implications and four applications of evolutionary and ecological interest. *Biol. J. Linn. Soc.* **10,** 229–49.

Sokal, R. R. and P. H. A. Sneath 1963. *Principles of numerical taxonomy*. San Francisco: W. H. Freeman.

Solbrig, O. T. 1972. New approaches to the study of disjunctions with special emphasis on the American amphitropical desert disjunctions. In *Taxonomy, phytogeography and evolution*, D. H. Valentine (ed.), 85–100. London & New York: Academic Press.

Soper, J. H. and F. H. Perring 1967. Data processing in the herbarium and museum. *Taxon* **16,** 13–19.

Sparks, B. W. and R. G. West 1972. *The ice age in Britain*. London: Methuen.

Squires, R. 1978. Conservation in Upper Teesdale: contributions from the palaeoecological record. *Trans Inst. Br. Geogrs* **3,** 129–50.

Stearn, W. T. 1973. *Botanical Latin*, 2nd edn. Newton Abbot: David & Charles.

Stebbins, G. L. 1950. *Variation and evolution in plants*. New York: Columbia University Press.

Stebbins, G. L. 1971. *Chromosomal evolution in higher plants*. London: Edward Arnold.

Stebbins, G. L. and A. Day 1967. Cytogenetic evidence for long continued stability in the genus *Plantago*. *Evolution* **21,** 409–28.

Stebbins, G. L. and J. Major 1965. Endemism and speciation in the California flora. *Ecol. Monogr.* **35,** 1–35.

Steenis, C. G. G. J. van 1950. The delimitation of Malaysia and its main plant geographical divisions. *Flora Malesiana* **1** (1), lxx–lxxv.

Steenis, C. G. G. J. van 1962. The land-bridge theory in botany. *Blumea* **11,** 235–542.

Steenis, C. G. G. J. van 1971. *Nothofagus*, key genus of plant geography in time and space, living and fossil, ecology and phylogeny. *Blumea* **19,** 65–98.

Steenis, C. G. G. J. van 1972. *Nothofagus*, key genus to plant geography. In *Taxonomy, phytogeography and evolution*, D. H. Valentine (ed.), 275–88. London & New York: Academic Press.

Steers, J. A. 1970. *An introduction to the study of map projections*, 15th edn. London: University of London Press.

Stoddart, D. R. 1977. Biogeography. *Prog. Phys. Geog.* **1,** 537–43.

Stoddart, D. R. 1978. Biogeography. *Prog. Phys. Geog.* **2**, 514–28.

Stromeyer, F. 1800. *Commentatio inauguralis sistens historiae vegetabilium geographicae specimen*. Göttingen: H. Dieterich.

Sukachev, V. and N. Dylis 1968. *Fundamentals of forest biogeocoenology* (trans. J. M. MacLennan). Edinburgh & London: Oliver & Boyd.

Sykes, W. R. and E. J. Godley 1968. Transoceanic dispersal in *Sophora* and other genera. *Nature* **218**, 495–6.

Szafer, W. 1975. *General plant geography* (trans. H. M. Massey). Springfield, Virginia: U.S. Dept. Commerce. (English trans. of *Ogólna geografia roślin*, 3rd edn, Warsaw: Państwowe Wydawnictwo Naukowe, 1964.)

Tarling, D. H. 1978. The geological–geophysical framework of ice ages. In *Climatic change*, J. Gribbin (ed.), 3–24. Cambridge: Cambridge University Press.

Tarling, D. H. and S. K. Runcorn (eds) 1972. *Implications of continental drift to the Earth sciences*. London & New York: Academic Press.

Tarling, D. H. and M. P. Tarling 1972. *Continental drift. A study of the Earth's moving surface*. Harmondsworth: Penguin. (First published by Bell 1971.)

Tansley, A. G. 1935. The use and abuse of vegetational concepts and terms. *Ecology* **16**, 284–307.

Tauber, H. 1967. Differential pollen dispersion and filtration. In *Quaternary paleoecology*, E. J. Cushing and H. E. Wright (eds), 131–41. New Haven, Connecticut: Yale University Press.

Thorne, R. F. 1972. Major disjunctions in the geographic ranges of seed plants. *Q. Rev. Biol.* **47**, 365–411.

Tivy, J. 1971. *Biogeography. A study of plants in the ecosphere*. Edinburgh & London: Oliver & Boyd.

Troll, C. 1939. Luftbildplan und ökologische Bodenforschung. *z.d. Ges. F. Erdk. zu Berlin* 1939, 241–98.

Tryon, R. 1971. Development and evolution of fern floras of oceanic islands. In *Adaptive aspects of insular evolution*, W. L. Stern (ed.), 54–62. Seattle: University of Washington Press.

Tsukada, M. 1966. Late Pleistocene vegetation and climate in Taiwan (Formosa). *Anthropology* **55**, 543–9.

Turrill, W. B. 1953. *Pioneer plant geography: the phytogeographical researches of Sir Joseph Dalton Hooker*. The Hague: Nijhoff.

Turrill, W. B. 1963. *Joseph Dalton Hooker. Botanist, explorer and administrator*. London: Nelson.

Tuzo Wilson, J. (ed.) 1972. *Continents adrift* (readings from *Scientific American*). San Francisco: W. H. Freeman.

United Nations 1972. *Declaration on the human environment*. Stockholm Conference, 1972.

Valentine, D. H. (ed.) 1972. *Taxonomy, phytogeography and evolution*. London & New York: Academic Press.

Van der Hammen, Th. 1974. The Pleistocene changes of vegetation and climate in tropical South America. *J. Biogeog.* **1**, 3–26.

Vavilov, N. I. 1927. Geographical regularities in the distribution of the genes of cultivated plants. *Bull. Appl. Bot. Genet. & Plant Breeding* **17**, 420–8. (Russian; English summary.)

Vavilov, N. I. 1941. *The origin, variation, immunity and breeding of cultivated plants*. Waltham, Mass.: Chronica Botanica.

Walker, D. (ed.) 1972. *Bridge and barrier: the natural and cultural history of Torres Strait.* Canberra: Australian National University.

Wallace, A. R. 1880. *Island life: or, the phenomena and causes of insular faunas and floras, including a revision and attempted solution of the problem of geological climates.* London: Macmillan (2nd edn 1892).

Walter, H. 1954. *Einführung in die Phytologie. III. Grundlägen der Pflanzenverbreitung. II Teil: Arealkunde.* Stuttgart: Ulmer.

Warming, E. 1909. *Oecology of plants. An introduction to the study of plant-communities* (trans. P. Groom and I. B. Balfour). Oxford: Clarendon Press.

Washburn, A. L. 1973. *Periglacial processes and environments.* London: Edward Arnold.

Watson, H. C. 1852. *Cybele Britannica,* Vol. 3. London: Longman.

Watson, H. C. 1873. *Topographical botany.* Thames Ditton: privately printed. (2nd edn 1883, London.)

Wattez, J.-R. 1968. *Contribution à l'étude de la végétation des marais arrière-littoraux de la plaine alluviale Picarde.* Doctoral thesis, Université de Lille.

Watts, D. 1971. *Principles of biogeography.* New York: McGraw-Hill.

Wegener, A. 1915. *Die Entstehung der Kontinente und Ozeane.* Braunschweig: F. Vieweg & Sohn. (2nd edn 1920, 3rd edn 1922, 4th edn 1929.)

Wegener, A. 1924. *The origin of continents and oceans* (trans. of 3rd German edn by J. G. A. Skerl). London: Methuen. (Trans. of 4th German edn by J. Biram, 1966. New York: Dover.)

Werner, Y. L. 1977. Manual mapping of locality records – an efficient method. *J. Biogeog.* **4,** 51–3.

West, R. G. 1977. *Pleistocene geology and biology,* 2nd edn. London: Longman.

Wettstein, R. von 1908. Untersuchungen über den Saison-dimorphismus im Pflanzenreiche. *Akad. Wiss. Wien. Denkschr.* **70,** 305–46.

Whitmore, T. C. 1969. First thoughts on species evolution in Malayan *Macaranga.* (Studies in *Macaranga III.*) *Biol. J. Linn. Soc.* **1,** 223–31.

Whitmore, T. C. 1975. *Tropical rain forests of the Far East.* Oxford: Oxford University Press.

Whittle, T. 1970. *The plant hunters.* London: Heinemann.

Willis, J. C. 1922. *Age and area. A study in geographical distribution and origin of species.* Cambridge: Cambridge University Press.

Willis, J. C. 1940. *The course of evolution by differentiation or divergent mutation rather than by selection.* Cambridge: Cambridge University Press.

Willis, J. C. 1949. *The birth and spread of plants.* Waltham, Mass.: Chronica Botanica.

Willis, J. C. 1973. *A dictionary of the flowering plants and ferns,* 8th edn, rev. H. K. Airy Shaw. Cambridge: Cambridge University Press.

Windley, B. F. 1977. *The evolving continents.* New York: John Wiley.

Wood, C. E. 1971. Some floristic relationships between the southern Appalachians and western North America. In *The distributional history of the biota of the southern Appalachians. 2. Flora,* P. C. Holdt (ed.), 331–404. (Res. Div. Monogr. **2**). Blacksburg: Virginia Polytechnic Institute and State University.

Wright, A. E. and F. Moseley (eds) 1975. *Ice ages: ancient and modern.* Liverpool: Seel House.

Wright, H. E. 1971. Late Quaternary vegetation history of North America. In *The Late Cenozoic glacial ages,* K. K. Turekian (ed.), 425–65. New Haven, Connecticut: Yale University Press.

Wulff, E. V. 1943. *An introduction to historical plant geography* (trans. from Russian by E.

Brissenden). Waltham, Mass.: Chronica Botanica.

Wulff, E. V. 1944. *Historical plant geography: history of the floras of the world.* Moscow: Akademiya Nauk SSSR (in Russian).

Wurdack, J. J. 1970. Certamen Melastomataceis. XV. *Phytologia* **20,** 369–89.

Zaleska, Z. 1950. *Metasequoia glyptostroboides* w świetle badań paleobotanicznych. *(Metasequoia glyptostroboides* in the light of palaeobotanical studies.) *Wiad. Muzeum Ziemi* **5.** Warsaw.

Index of plant names

Numbers printed in *italics* refer to text figures.

★For authority citations relating to the supra-generic taxa of the Proteaceae mentioned in Table 7.1, see Johnson and Briggs (1975, 170–4).

Virotia L. Johnson & B. Briggs 103
Viscum album L. 106

water crowfoot 5
water fir 112, *113*
water-lotus genus *96,* 97
water mint 20
white rockrose 88

white spruce 109
willow *83*
Winteraceae Lindl. 74, 107
witch-hazel *92*
wood dog violet *29*

Zingiber Boehm. 106
Zostera L. 106

Subject index

Numbers printed in *italics* refer to text figures.